Cataclysm by Capitalism

P.R. Brewbaker

Cataclysm by Capitalism

ISBN
978-1-387-02608-1

Printed in the United States of America
By
www.lulu.com

Table of Contents

Introduction

In 1972, *The Limits to Growth* study addressed how humans would adapt to the physical limitations of planet Earth. It forecast that from the year 2000 to 2050 the ongoing growth in the human population would stop-either through catastrophic "overshoot and collapse" or a well-managed "peak and decline."

In this book we look at the root causes of our needs for a high rate of consumption. As a "successful" species with unique talents we have failed to recognize, even to this day, that we need to adopt a "big picture" approach of looking at Earth and its biota.

It is imperative that we slowdown from our daily routines and recognize ourselves as part of and not separate from this beautiful life giving planet. We need to focus on human well-being rather than on per capita income growth. We will be constrained into unfortunate situations in surprising ways if we do nothing.

So, how do we prepare for the years ahead? With heart, fact, and wisdom, I will guide you along a realistic path from the distant past into the future and discuss what readers can do to ensure a better life for themselves and their children during the increasing turmoil of the next 30 years.

Presently, we need to look at life, its origins and its modes-operandi.

Preamble

May the Cycle be Unbroken

In the beginning, 4 billion years ago, there was a dance on Earth to the laws of physics by rock, wind, and water powered by the rays of the sun. A person could say the Earth was alive with rain falling, rivers flowing, and landscapes transforming: all dancing in cycles. Yet to Mother Earth, it was not alive enough, and so new dance cycles came into being 3.8 billion years ago guided by genetic laws of the organic matter. As the number of organic dance cycles grew, they started to interlock, forming circular dancing chains that too interconnected: a kaleidoscope of colors and forms. This state of affairs transformed Earth's surface into stages of green for the dramas of the song of birds, the roar of bull elephants, and the discourse of tribal leaders to name just a few and far between.

The fruits of one dance cycle were given freely to other dance cycles that then recycled them back to Earth. Again, sunlight fueled these new dance cycles. Each new organic life form grew through the stages of thriving, expanding, acclimating, and maturing hand in hand with each other and with the finite Earth. Each thrived on the available resources with added efficiency, expanded its territory when it had the chance, adjusted its lifestyle by finding its ideal habitat, and matured in a way that added stability to its environment by teaching its children to dance to the circular rhythm of the Sun–Earth beat.

In a blink of an eye 10,000 years ago, dance cycles started to break. Chains unraveled, stages collapsed, and colors faded. "Why, oh why was this happening?" exclaimed Mother Earth as she cried and searched for the answer. Finally, she discovered a species that was stuck in the expansion stage and could not get to the stage of maturation to continue the cyclical dance. This species developed endless self-gratification to the point it could not see how it was detrimentally affecting its dance and other dance cycles and breaking up the fabric of life on Earth. In many instances, the dance of this species stopped the dance of others who then vanished.

When Mother Earth questioned this species on their behavior, it responded, "We have the right to be free and to promote freedom around the world. No one is going to tell us what to do." It did not understand that it needed to continue growth through all dance stages. Maturation added stability to Earth for its children and the children of others. When questioned on that point it only came back with, "I can manage my own dance and the dance of others on Earth just fine through the use science and technology. I can do it. Just look at my achievements. I have conquered the Earth and even the moon!"

The Earth was powerless to change this immature species' mind. After all, they originated from her, and she knew she could not change herself.

P.R. Brewbaker

The Way of the Earth and Sun

Informally, the origin of life or abiogenesis is the natural process by which life arises from nonliving matter, such as simple organic or inorganic compounds. It is thought to have occurred on Earth between 3.8 and 4.1 billion years ago. Abiogenesis is studied through a combination of laboratory experiments and extrapolations from the characteristics of modern organisms, and aims to determine how pre-life chemical reactions gave rise to life on Earth.

In simple terms, life is a chemical reaction requiring energy. With a fixed amount of incoming energy from the Sun (approximately 3,459,000 calories per square yard) hitting the Earth's surface, lowering the amount of energy required for this chemical reaction to take place is what evolution is all about. If you ask how does lowering the amount of energy in a chemical reaction is achieved, the answer is in the availability of a catalyst. In most cases where a catalyst is required, reactions occur faster because they require less activation energy (Please refer to Appendix 1 for more information on catalysts and evolution).

In summary, evolution is self-replication of catalysts where a change or changes occur to the offspring that lead to more energy efficient catalysis.

Cycles

For the most part, Earth's matter moves in a circular fashion. For example, water in the ocean evaporates due to the heat of the sun and forms clouds that may travel to exposed lands. There, these clouds burst into rain creating streams of water that aggregate to create creeks and rivers that carry the water once again back to the ocean. This is an example of the hydrologic cycle of water on Earth.

Such cycles are the rule and not an exception for the movement of matter on Earth and are manifested in each species with a circulatory system. Groups of species such as those found in a forest circulate matter in the same way. One can say that energy drives the circulatory system of forests on Earth, and that without this circulatory system, forests would collapse because each of its links, be it another species or inanimate matter, is a tiny step in recycling of life-giving nutrients, making them available for regeneration of life and therefore maintaining the life of the forest.

When we look out a window or stroll through a park, we may see trees, birds, grass, and bees to name just a few of the life forms present. If we focus on a flower, many of us look at it as an object independent of the other objects around it, but in reality, it is intimately connected with all other objects in the picture we see.

Imagine that the flower we are looking at is clover. Clover needs soil to grow in, and as a mature plant, it provides nourishment for the bison roaming the Midwest plains. The puma on the other hand seeks the bison for food and when it manages to bring one down, it drags what it can of the carcass to the shade of a tree to eat. Flies gather around the carcass to feed and lay their eggs avoiding the birds that feed on them.

The above pictures and explanation are simplified, but they show how different species are interlinked. A more realistic representation of the link between species is circular as shown in the following graph.

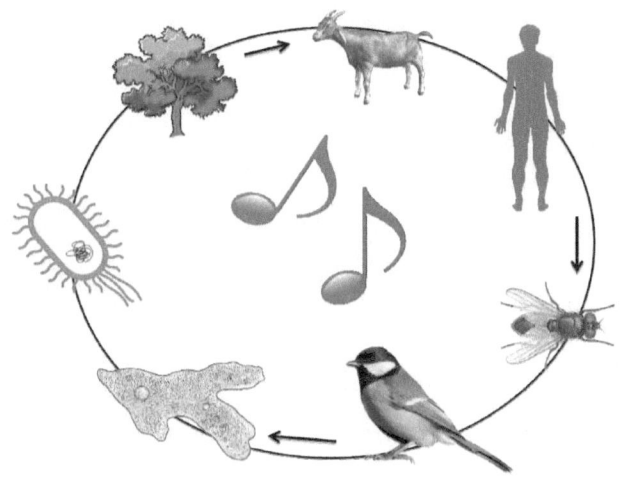

If we follow the biological cycle above clockwise starting with the green bush at the top left, we can tell a story of how nutrients flow from one live object to another. First, the bush extracts water and minerals from the soil and with the help of the sun and atmosphere produces tender leaves which are consumed by the goat to build its body and maintain its growth and health and to give birth to its offspring. The goat in turn is consumed by the human in the picture. The leftovers from the goat are used by the fly to lay its eggs. Some birds eat flies, and all of the preceding species—the goat, human, bird, and fly—as a result of digestion, leave behind fecal matter that is consumed by various microbes such as amoebas and bacteria, which return the nutrients (extracted from the soil by the green bush) to the soil once again.

Like the laws of physics that govern the universe in its grand symphony of reality, the musical notation in the preceding diagram of the living cycle signifies the rhythm with which this process is carried out on Earth.

On any territory, there are multitudes of living cycles but with different species that participate in returning the nutrients absorbed and transformed by the plants back to the soil and to the atmosphere. In addition, these living cycles can have relations with one another. For instance, a living cycle that starts with a bush can have more than one grazer consuming it such as goats, sheep, and deer. These living cycles can take the following shape:

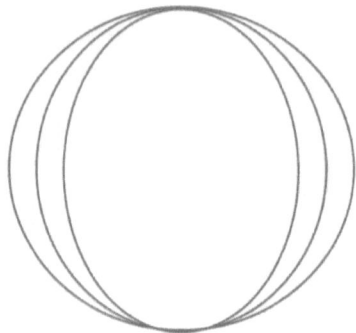

If we look at a square acre of land and try to describe all the living cycles within it, we may come up with a million combinations, some of which may look like the living cycles in the next diagram:

Living cycles are not limited to sources of food. They can be related to nest building and so on. The following biological cycle graphic depicts a bird building a nest in a tree. When the eggs hatch, the nest may fall to the ground where insects like termites break it down passing on the resulting matter to amoebas and bacteria resulting in the return of the nutrients back to the soil.

This is probably a good point to introduce the concept of an ecosystem. A forest is an example of one. An ecosystem is a mega catalyst that describes the relations or circular links between air, soil, water, plants, and animals of a forest. A forest is not the only ecosystem. A desert such as the Sahara is also considered an ecosystem as well as the biota of a continent.

Cycle Stability

At this point, with respect to the living cycle, one might argue that this cycle is unstable. For example, a gazelle herd can consume all the vegetation in their territory then starve to death the next week or year. At the same time the living cycle of the gazelle also dies because moving clockwise in the following diagram, all subsequent living things will also starve.

But we know that gazelle graze off of the vegetation in their territories without consuming it all. We may ask the question, "How do they temper their appetites so that they do not exhaust the vegetation?"

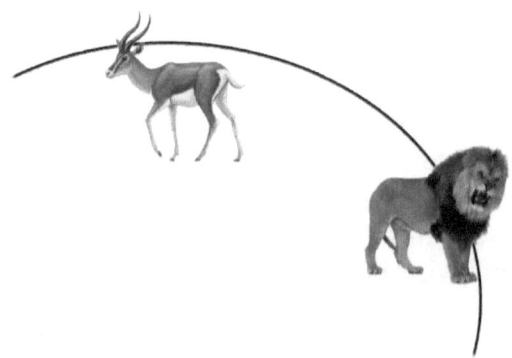

When we look at the relationship between the lion and gazelle, the African lioness eats from the herd of gazelle to satisfy her hunger and the hunger of her pride. She does this in a way that always leaves enough gazelle for her to hunt the next day or the next year. If she finished off the whole herd of gazelle, then the lioness'

cubs will have nothing to eat the next day or year and die off, the end result being a broken living cycle.

One key factor that gives stability to this living cycle is laziness. In the wild, it has been observed that lions sleep a lot under the shade of trees and go out hunting for gazelle or other prey only when their bellies rumble; to conserve energy lions go after slow individuals in the gazelle herd. If laziness was not present in the pride of lions, one can imagine the lionesses hunting all day and bringing down as many gazelle as they could, which would in the end break the living cycle.

Another key factor is territoriality. Lions are known to have territories that keep other hungry lion prides out and therefore preserve the herd of gazelle from overhunting. At the same time, the gazelle are involved in the preservation process too. They have evolved the ability to flee when the hunt is on. . In addition the gazelle have a mechanism that keeps them from over grazing.

Thus, with the lion helping to control the number of individuals in the herd of gazelle, the gazelle in turn tend not to overconsume its food source. All these physical and behavioral characteristics are genetically controlled and evolve to help stabilize the living cycle.

Another factor in living-cycle stability is the availability of substitute species. For example, if a fly—that uses the lion pride's leftovers of food as a habitat for their larvae—goes extinct, then another species will take over as a substitute in processing these leftovers so that the living cycle is not broken.

Then there are the generalist species of flies that use leftovers from several other predator species. For instance, in an ecosystem, there may be a fly that can lay eggs on leftover carcasses of prey of the lion, cheetah, hyena, or wild dogs.

In summary, living cycles are stable over time because the genetics of their participants have adapted to self-preservation in the course of evolution.

Cycle Instability: The Algae Bloom

The term "bloom" refers here to the explosion of algae density in a pond. To understand what an algae bloom does, we first must understand how a pond of water operates. A pond naturally occurring in a forest is composed of a pond bottom consisting of soil or clay that is a depression in the ground filled with water. On top of the water is air. Algae grow in water and receive sunlight from above and nutrients from the water and the soil beneath. As the algae grow, mature, and die, they sink to the bottom of the pond. There, bacteria decompose the algal matter and release key elements that then circulate in the pond water nourishing subsequent generations of algae.

Algal blooms in ponds are common. They occur when the conditions are right for algae to undergo a sudden or prolonged burst of growth. This burst is usually due to an increase in one or more chemical resources in the water essential for algal growth such as compounds containing phosphate.

Most blooms are benign toward the water system. The water may turn bright green for a few days or a week or so. The algae that bloomed die and sink to the bottom of the pond when the chemical or chemicals that initiated the bloom are exhausted.

Some water systems experience long blooms. These may last from a month to two or longer so long as the chemical resources responsible for the bloom continue to come into the pond water system, such as an effluent from sewer systems. This process can be devastating to the water system's biota and cycles because the increase in the number of alga corpses being consumed by bacteria (which are oxygen consumers) will cause the lake to become oxygen deficient for many plants and animal species, which in turn will die off. This type of bloom chokes life out of the lake, which will eventually die, culminating in anaerobic bacteria's jumping into action, producing foul smells like rotten eggs associated with stagnant water systems, after most if not all oxygen is consumed.

You would think that even if the chemical sources that triggered the bloom in the first place continue to flow into the pond, the bloom will continue indefinitely, but on the contrary, the blooming species will eventually die off and sink to the bottom of the water system if other chemical resources necessary for life are depleted as a result of the bloom. You see, algae depend on many chemical resources present in the water system and when one or more of these are depleted, the bloom cannot continue, and the species in question dies off sinking to the bottom, where the bacteria go to work.

The density of algae in a pond undergoing a long bloom

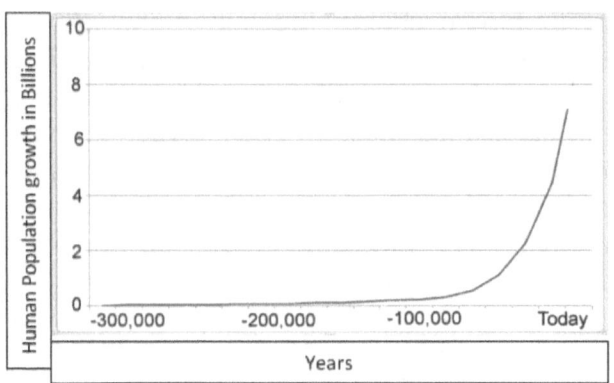

Is Earth Undergoing a Human Bloom?

To bring back a pond that has experienced a bloom, we need to take out the algae, plant matter, and animals that died during the bloom and the associated decomposition-related bacteria. In addition, all sources of chemicals responsible for the bloom must be stopped from entering the pond. Only then will enough oxygen be reabsorbed by the pond from the air above, allowing fish, plants, and algae to grow once again and continue the life cycle of the pond.

You may ask, "How does the explanation of pond blooms relate to the formation and stability of cycles?" Mankind is believed to be undergoing a "bloom" that carries with it a mass extinction of 50% or more of all life on Earth. Underlying this argument is the idea that we as humans and our habitat are no different from the algae in the pond. It's all a matter of chemical reactions that follow physical laws over which we apparently have no control.

In these modern times, we face a whole new set of challenges that differ from the challenges faced by our Greek ancestors such as Socrates and Plato. In those early times, the challenges were those of aggregating people together into cities that are under the control of fair and acceptable governing bodies and laws. Today, we face a whole new set of problems related to the biological "success" of our species.

The Niche

The niche of the fly is everything necessary for the survival of the fly that is not the fly itself. For example, the image to the left above is of the fly and not its niche. The gray image to the right of the fly is everything else on which the fly depends minus the fly itself and is called its niche. The gray area is the air the fly depends on for flight as well as breathing, the food that it requires, the material in which it lays its eggs, the temperature of the surroundings where it goes to sleep at night, the perch on which it stands, the fly that fertilizes its eggs, and so on.

The niche is also the environment in which the fly lives most efficiently. You could say that the niche is a perfectly tailored suit for the fly. For example, a fly that is used to live in the tropics will die in the polar areas because of the temperature. A tropical fly that is transported to a temperate region will not survive either because it will not find the right food or shelter. A tropical fly that is moved from the tropics of Africa to the tropics of Central America will probably have a harder time being as efficient as the local African flies.

Therefore, the suitability of a niche is important to the species and in turn to the living cycle in which it serves as a conduit of energy and other resources. For instance, in the diagram above, if the environment changes rendering the bird's niche harder to live in, the birds will not be as efficient at controlling the population of flies, and therefore the flies will multiply faster, and in the end, be a nuisance for the gazelle and lions, who will in turn waste more energy and time trying to chase them away or avoid the fly hoards, to say nothing of the fly's ability to spread a disease disrupting the living cycle.

To recap, the niche comprises all the resources required to fulfill an individual's livelihood and role in the balance of the living cycle. The niche of the gazelle is a part of the niche of the lion, and the niche of the lion is a part of the niche of the fly, and so on.

Succession of Mega Catalytic Ecosystems

An ecosystem is a collection of species that are interdependent on each other and on the nonliving surroundings in which they evolved such as mountains, soil, water, and energy. We can also call an ecosystem a unique mega catalytic organism because its cooperating individuals include large numbers or plant and animal species. Ecosystems go through a process of environmental succession. This succession is driven by the ever-changing conditions on Earth such as climate and terrain. For example, with respect to climate, there were ice ages at different periods in the ancient past. Some territories were once deserts that are now lush forests and vice versa. As for terrain, whole continents are constantly moving around the globe, volcanoes belch up magma from deep underground, rain pulverizes rocks converting them to soil, and so do certain plants.

These changes in turn alter the relationships between species as well as the types of species present in an ecosystem. For example, if we start with a rocky terrain we often find that the living cycles present are simple ones and involve miniature species. If you look at a rock in a temperate zone of the globe, you may find more often than not lichen growing on it. Like any species, lichen grows and dies. Its remains are then consumed by various insects and microbes, and the nutrients it was able to absorb from small cracks in the rock are released into the environment once again.

As the lichen grows and creates more cracks in the rock's surface, the rock starts to flake away, and the flakes accumulate creating soil that other plants, such as thistles and legumes, can use to grow. This process creates a new type of living cycle that involves herbivores like the bison, carnivores like the puma, and parasites like flies. Not only are the species involved larger in size than those in the lichen living cycle, the number of species involved also grows. For example, the following pictures show this transition from left to right of a simple living cycle to a more complex one. This is called ecosystem succession.

You should realize that the start of the living cycles above are the plants that take up minerals and water from the soil, carbon dioxide from air, and energy from the sun, thus producing living matter and in turn feeding the other species in each living cycle. These plants also create whole three-dimensional environments above ground as the succession process continues into forests and jungles containing thousands of living cycles and millions of species.

The new species that fill these new ecosystems are generated by evolution.

Maturation

Evolution operates on a trial-and-error basis. Those born with an error or sub performing trait perish, whereas those born with a successful trait survive. So, it is with the relationships of species that belong to a cycle. If a species does not fit properly in its cycle, then it will die and be replaced by one that is a better fit. This process continues until the cycle is most stable.

For example, when a mezzo catalytic African lion consumes a gazelle, it squeezes most of the energy out of the gazelle's flesh to maintain itself, produce young, and then it leaves the remainder of its digestion on the ground. This remainder is then consumed by insects, insect larva, and a host of microbes that squeeze even more energy out of what that lion left behind. This process releases critical minerals into the ground, making them available to the grass and legume seeds and plants that in turn use them to convert the sun's energy into lush vegetation.

This vegetation is then consumed by the gazelle and the cycle continues, the end result being zero impact, where a few critical minerals are transported up the stalks of grasses and legumes with the help of the sun, then down to the ground once again through the gazelle, lion, insects, and microbes.

Lions spend much of their time resting and are inactive for ~20 hours a day. Although lions can be active at any time, their activity generally peaks after dusk with a period of socializing and grooming. Intermittent bursts of activity follow through the night hours until dawn, when hunting most often takes place. They spend an average of two hours a day walking their territories and 50 minutes eating.

The size of adult lions varies across their range. In Zimbabwe, males average approximately 418 lbs. (280 lbs. for females), while in East Africa, males average 386 lbs. and females 263 lbs.

Lionesses do most of the hunting for their pride. They are more effective hunters because they are smaller, swifter, and more agile than the males and unencumbered by the heavy and conspicuous mane, which causes overheating during exertion. They act as a coordinated group with members performing the same role consistently in order to stalk and bring down the prey successfully. Smaller prey is eaten at the site of the hunt, thus being shared among the hunters; when the kill is larger, it often is dragged to the pride's area. There is more sharing of larger kills although pride members often behave aggressively toward one another as each tries to consume as much food as possible. Near the conclusion of the hunt, males have a tendency to dominate the kill once the lionesses have succeeded. They are more likely to share this food with the cubs than with the lionesses, but males rarely share food they have killed by themselves.

The African lioness eats from the herd of gazelle to satisfy her hunger and the hunger of her pride. She does this in a way that always leaves gazelle for her to hunt the next day or the next year. You might ask, "How does she do that?" The genetics of the lion has matured over time and ensures conservation; if a lioness finished off the whole herd of gazelle, then her cubs would have nothing to eat the next day or year and die off, the pride of lions would go extinct.

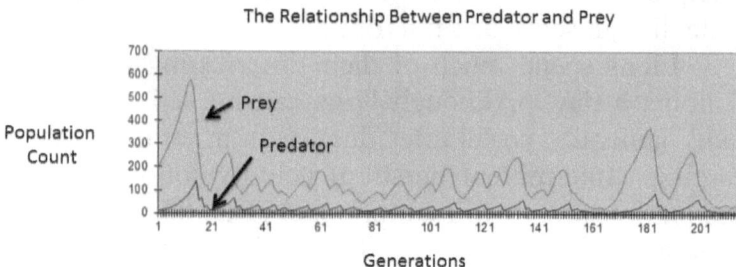

The above graph shows a stable relationship between a predator and its prey over 210 generations. The fluctuation in the population count accounts for perturbations in the environment such as years when the foliage that the gazelle feed on is scarce or abundant.

Summary

There should be no doubt that any one catalytic organism is connected in some way or another to every other creature in an ecosystem through the cascade of sun's energy from plants to the rest of the biota. It should also be clear that no one creature or cycle is more important than another in an ecosystem community. In addition, the sun's energy is a limited resource, and life forms, strive to use it in the most efficient way possible.

The Emergence of Man

Human biological evolution is the process of change by which people originated from apelike ancestors. One of the earliest human defining traits is the ability to walk on two legs, which evolved over 4 million years ago. Other important human characteristics such as a large and complex brain and the capacity for language developed more recently. Many advanced traits including complex symbolic expression, art, and elaborate cultural diversity emerged mainly during the past 100,000 years.

The evolution of bipedalism, a larger brain, and the capacity for language all contributed to the efficiency with which these early human catalysts acquired the scarce energy needed to survive.

Humans are primates. Physical and genetic similarities show that the modern human species, *Homo sapiens*, has a very close relation to another group of primate species, the apes. Humans and the great apes of Africa (chimpanzees, baboons, and gorillas) share a common ancestor that lived between 8 and 6 million years ago. Humans first evolved in Africa, and much of human evolution took place on that continent. The fossils of early humans who lived between 6 and 2 million years ago come entirely from Africa.

Currently, approximately 20 species of early humans are believed to have existed. Scientists do not know how some of these human species are related. Many early human species left no living descendants. Scientists also debate how to identify and classify particular species of early humans, and what factors influenced the evolution and extinction of each.

More important is the fact that humans like all the life forms on Earth owe their lineage to the very first living entity that must have appeared on Earth 3 or 4 billion years ago. This lineage carries with it the ability to use cooperation, symbiosis, parasitism, and a host of other traits to get established in a niche belonging to a cycle or cycles in an ecosystem.

This human niche was one of many participants in the efficient use of the sun's limited energy that reached the Earth where its capture started with plants and similar organisms. Once captured, it cascaded down through each ecological cycle from a prey to predator, from a host to parasite, or was shared in a symbiotic relationship to name just a few.

Hunter-Gatherers

Over 3 million years ago, our ancestors began using tools made of stone and bone that seem to be associated with hunting and digging. This early hunting and gathering played a big part in humans' evolving from *H. erectus*, who had spread throughout Africa and then Europe a million years ago, to eventually become *Homo sapiens*, who subsequently spread throughout the world. Beginning around 10,000 years ago, humans began domestication of animals and plants, becoming farmers and herders. The human lifestyle changed dramatically, with towns and cities and empires developing quickly. By 1500 A.D., at the dawn of European expansion, hunter-gatherers still occupied almost one-third of the world's landmass, including all Australia, the northwestern half of North America, and the southern part of South America, including parts of sub-Saharan Africa and parts of Asia. Over time, hunter-gatherers were able to survive with their cultures intact only in areas that were of little interest to the rest of humanity, such as places with little water, thick jungles, impassible mountains, and frozen Arctic areas. In addition, attempts to study these so-called primitive tribes have been lethal, where up to 50% of their populations die of so-called modern ordinary diseases such as the flu.

The Organization of Hunter-Gatherer Groups

The hunter-gatherer lifestyle lasted for 3 million years and proved itself to be at zero ecological impact status by surviving that way for so long without damaging the environments. Some hunter-gatherers still exist today hiding in various inaccessible geographic regions. These hunter-gatherer societies are typically organized by having small bands of ~30 individuals that are quite independent, but connected to a larger tribe of several or many similar groups. Individuals may move from one band to another, and often young adults will select their spouses from other bands within the larger tribe. These bands will usually move from place to place in a seminomadic way although some may remain in permanent settlements. Each band has a leader, or chief, who usually is an organizer and mediator, but has little power beyond that. These groups hardly possess any stockpiles of material wealth, and leaders particularly avoid having any more possessions than other members of the group do, so that there will not be any resentment or jealousy in the group. Accordingly, a chief usually does not have much power over the rest of the members of the group. All members of the band have an equal voice in decision making. The chief only organizes those voices. In fact, if one person becomes too aggravating to another person of the tribe, that person can simply move to another band. The chief is always under pressure to be humble and accommodating because he does not want migration of people away from his band.

The greater complexity of food-producing groups as compared to those who hunt and gather is evident in their types of leadership. Hunter-gatherers typically live in bands without a central leader. Instead of a ruler, leadership is in the form of respected individuals who can persuade the group to do things, but who have no real power. Given that everyone has a relatively equal share of the available resources, hunter-gatherers are egalitarian. In contrast, horticulturalists and pastoralists generally live in tribes, which are larger, more organized groups either without a central leader or in chiefdoms in which one individual has absolute power that is inherited. There must be a surplus of food—generally obtainable only through agriculture—for a chiefdom to exist because the chief increases his control by collecting the surplus and redistributing it. The hierarchy of power means that there is less equality in food-producing societies (as opposed to hunter-gatherers), so that resources are not always evenly shared. Agriculture provides the circumstances where disparities of wealth among households are possible, and the wealthier households begin to exploit poorer units.

Cave Paintings

These are paintings found on cave walls and ceilings, and especially those of prehistoric origin, which date back to some 40,000 years ago in both Asia and Europe. The exact purpose of the

Paleolithic cave paintings is not known. Evidence suggests that they were not merely decorations of living areas because the caves in which they have been found do not have signs of ongoing habitation. They are also often located in areas of caves that are not easily accessible. Some theories hold that cave paintings may have been a way of communicating with others, whereas other theories ascribe a religious or ceremonial purpose to them. The paintings are remarkably similar around the world, with animals being common subjects that yield the most impressive images. Humans mainly appear as images of hands, mostly hand stencils made by blowing pigment on a hand held to the wall.

The most common subjects in cave paintings are large wild animals, such as bison, horses, aurochs, and deer, and tracings of human hands as well as abstract patterns, called finger flutings. The species found most often were suitable for hunting by humans, but were not necessarily the actual typical prey found in the associated deposits of bones; for example, the painters of a site in France have mainly left reindeer bones, but this species does not appear at all in the cave paintings, where equine species are the most common. Drawings of humans were rare and are usually schematic as opposed to the more detailed and naturalistic images of animal subjects. One explanation for this pattern may be that realistically painting the human form was forbidden by a powerful religious taboo. Other geologists suggest that climate controlled the themes depicted. Pigments used include red and yellow ochre, hematite, manganese oxide, and charcoal. Sometimes, the silhouette of an animal was incised in the rock first and in some caves all or many of the images are only engraved in this fashion.

Similarly, large animals are the most common subjects in the many small carved and engraved bone or ivory and stone pieces dating from the same periods. But these include the group of Venus figurines, which have no real equivalent in cave paintings.

Hand stencils, made by placing a hand on the wall and blowing pigment at it, form a characteristic image of a roughly round area of solid pigment with the uncolored shape of the hand in the center, which may then be decorated with lines or dashes. These are often found in the same caves as other paintings, or may be the only form of painting at a location. Some walls contain many hand stencils. Similar hands are also painted in the usual fashion. A number of hands show a finger wholly or partly missing. Some interpreted the paintings as being hunting magic, meant to increase the number of animals.

Another theory, developed on the basis of ethnographic studies of contemporary hunter-gatherer societies, is that the paintings were made by Paleolithic shamans. A shaman would retreat into the darkness of the caves, enter a trance state, and then paint images of his or her visions, perhaps with some notion of drawing power out of the cave walls themselves.

Studies of both highly artistic and publicized paintings and a variety of lower-quality art and figurines identify a wide range of skill and ages among the artists. There is a hypothesis that the main themes in the paintings (and other artifacts) such as powerful beasts, risky hunting scenes, and the representation of women in the Venus figurines are the work of adolescent males, who constituted a large part of the human population at the time. Nonetheless, after analysis of hand prints and stencils in French and Spanish caves, it has been proposed that a proportion of them are female hands.

Summary

Our original zero ecological impact niche was apparently one of hunter-gatherer nomadic small groups consisting of several families that foraged on fruits, berries, mushrooms, some grains, flowers, and game. Hunter-gatherers move on when these food sources reached a stage of reduction but were still able to recover.

These small groups had territories that were defended fiercely because the territories were needed for survival as small groups and did not represent an object of wealth or status. The territory protected their most simple of natural cyclical lifestyles.

Tools were made of local perishable materials consisting of tree branches, roots, animal skins, animal bones, and stones. The process of making these tools also involved perishable and local materials to guarantee their return to nature. The same goes for the construction of dwellings and shelters.

They had a symbolic language and therefore thought, as demonstrated by the art found in caves and by contemporary hunter-gatherer groups. Today's hunter-gatherer groups are healthy and free of modern human diseases such as obesity, hypertension, diabetes, and communicable diseases.

They had music and song in their lives that told stories of the universe, the hunt as well as those of love and appreciation. They had rituals for preserving the group and the environment. You may think of this as their religious demonstration.

The context of the hunter-gatherer lifestyle within its environment is the one that best describes the natural human niche that is in balance with nature supporting the zero ecological impact way of life through the cyclical character of life on Earth.

P.R. Brewbaker

The Agriculturalists

Agriculture began independently in different parts of the globe, and included a diverse range of taxa. At least 11 separate regions of the Old and New World were involved as independent centers of origin.

	Years Ago
Wild grains were collected and eaten from at least	105,000
Pigs were domesticated in Mesopotamia around	15,000
Rice was domesticated in China between	13,500 and 8,200
Sheep were domesticated in Mesopotamia between	13,000 and 11,000
Emmer and einkorn wheat, barley, peas, lentils, bitter vetch, chick peas and flax were cultivated in the Levant	11,500
Cattle were domesticated from the wild aurochs in the areas of modern Turkey and Pakistan some	10,500
In the Andes of South America, the potato was domesticated along with beans, coca, llamas, alpacas, and guinea pigs	10,000 and 7,000
Sugarcane was domesticated in New Guinea around	9,000
Sorghum was domesticated in the Sahel region of Africa by	7,000
Cotton was domesticated in Peru by	5,600
In Mesoamerica, wild teosinte was domesticated to maize by	6,000

In the Middle Ages, both in the Islamic world and in Europe, agriculture was transformed with improved techniques and the diffusion of crop plants, including the introduction of sugar, rice and cotton along with fruit trees such as the orange to Europe by way of Al-Andalus.

After 1492, the Columbian exchange brought New World crops such as maize, potatoes, sweet potatoes and manioc to Europe, and Old World crops such as wheat, barley, rice and turnips, and livestock including horses, cattle, sheep and goats to the Americas.

Irrigation, crop rotation, and fertilizers were introduced soon after the Neolithic Revolution and developed much further in the past 200 years, starting with the British Agricultural Revolution. Since 1900, agriculture in the developed nations, and to a lesser extent in the developing world, has seen large rises in productivity as human labor has been replaced by mechanization, and assisted by synthetic fertilizers, pesticides, and selective breeding.

The Haber-Bosch method allowed the synthesis of ammonium nitrate fertilizer on an industrial scale, greatly increasing crop yields. Modern agriculture has raised political issues including water pollution, biofuels, genetically modified organisms, tariffs and farm subsidies, leading to alternative approaches such as the organic movement and regenerative agriculture.

The Transformation from Hunter Gatherer

As you can imagine, cultivating plants required a more sedentary existence than that of the hunter gatherer nomadic groups. The structure of the group didn't change much where there is a leader and followers. As the groups got better at cultivation and animal husbandry a surplus developed. This surplus became possessions of the tribe distributed according to the guidance of the leader. Who is to get this surplus created the first political power struggles and power shifted from the group to the leader.

Territories were more firmly defended as the groups increased in its agricultural activities. Barter within the group and with other neighboring groups increased with possessions being a type of monetary entity where each possession had a value. For example, a cow may have had the value of five sheep to one group that was hungry for a taste of sheep meat and wool for spinning thread.

There were instances where one group needed maize in exchange for wheat. Since the harvest season for these crops is different and storage was no guarantee against rodents consuming the goods it was difficult to make such transactions. What was needed was something to hold the value of the wheat until the maize was ready to be harvested and exchanged. This is where a type of medium object was used to hold the value of the wheat until the transaction could be completed. The medium was exchanged for the wheat and returned when the maize was ready.

Many things have been used as medium of exchange, for example, livestock and sacks of cereal grain (from which the Shekel is derived) – things directly useful in themselves, but also sometimes merely attractive items such as cowry shells or beads were exchanged for more useful commodities. Precious metals, from which early coins were made, fall into both categories.

We leave this chapter as techniques of agriculture improved and so did its surplus. In Europe, small groups and tribes with their leaders and followers were replaced by feudal landlords and serfs but still dependent on the land for a living.

The Alien Methods: Capitalism

The feudal landlord and serf system started to give way to urbanization in Europe approximately in 12th century A.D. Towns were built and roads had been laid to connect them together and to the land. As towns grew into cities merchant capitalism developed which involved capital resources used to hire people who were willing to exchange labor for monies in working the land, providing a service or handmade products. Merchant capitalism also included renting of land or products as well as lending monies.

Sir Francis Bacon in the 1600s, among others, advocated the use of the scientific method in research. This ushered in the industrial revolution in the 1800s with the invention of machines that were driven by readily available fossil fuels replacing primitive biomass dependent services and hand made products.

In essence, the feudal system of landlord and serf was replaced by two human organizations in the industrial capitalistic age. The first was the town or city such as London where the masses needed goods and second was the factories that used paid laborers to provide these goods.

As far as the factory was concerned, capitalism was the production of goods in exchange for monies that was driven by the desire for personal accumulation of wealth in a free market system where the needs and wants of consumers are the driving forces of such markets.

In addition, capitalism as an economic system and mode of production can be summarized by the following:

- **Private ownership of the means of production.**
- **The investment of money to make a profit.**
- **Capital accumulation: Production for profit and accumulation of wealth as the implicit purpose of all or most of production organizations and the elimination of production formerly carried out on a common social or private household basis.**
- **Commodity production: Production for the exchange of goods for monies based on maximizing exchange value.**

The inhabitants of these towns and cities recognized that capitalism had a dangerous aspect in the free market equation namely the formation of monopolies that did more harm than good to the consumers. Laws were introduced to prevent such monopolies and encourage competition. This competition introduced the drive for innovation in the form of cost reduction, substitute and new product introduction technologies.

A monopoly exists when a specific person or enterprise is the only supplier of a particular commodity. Monopolies are thus characterized by a lack of economic competition to produce the good or service, a lack of viable substitute goods, and the possibility of a high monopoly price well above the seller's marginal cost that leads to a high monopoly profit. In economics, a monopoly is a single seller. In law, a monopoly is a business entity that has significant market power, that is, the power to charge overly high prices.

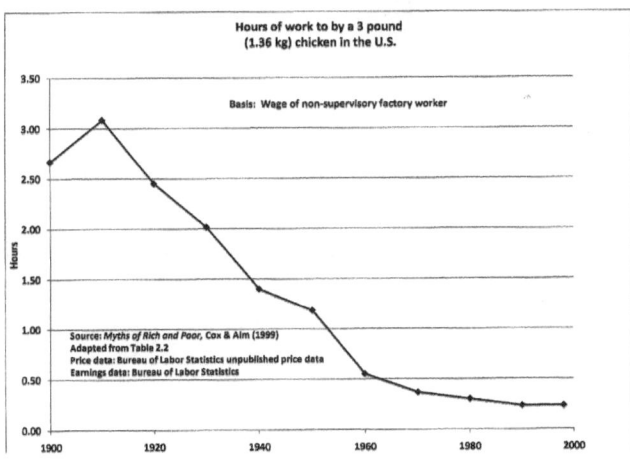

With the advent of capitalism and competition a slight increase in global population level can be detected during the 1500 to 1800s A.D. as shown in the following chart. The discovery and availability of fossil fuels became abundant through technological advances brought on by competition in the capitalistic market system resulting in the exponential growth of Earth's population from approximately 800 million in 1700 A.D. to 7.4 billion today reminiscent of the algae bloom introduced earlier in this publication.

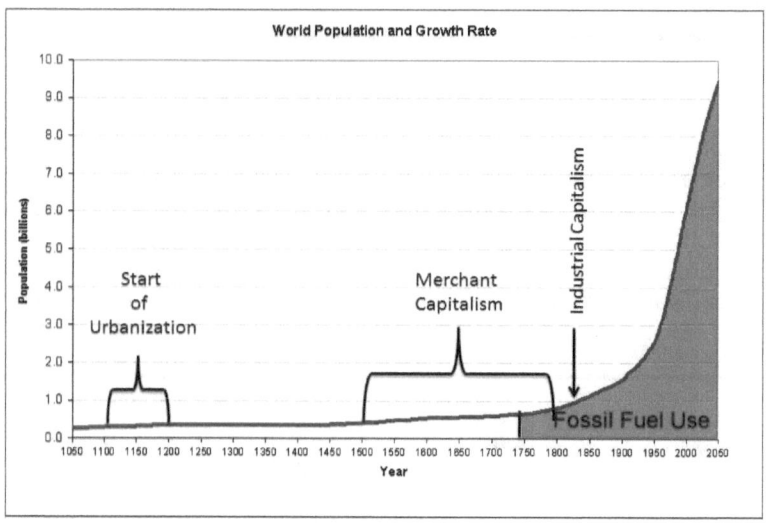

(For a more detailed history on the advent of industrial capitalism please refer to Appendixes 2 and 3)

With the increase of population count is an increase in the consumption of resources extracted from the Earth to the point that some are being depleted beyond economic value such as critical minerals as can be observed in the following table for example.

Mineral	Years Remaining
Silver	15
Lead	17
Zinc	20
Copper	22
Tin	28
Nickel	30
Iron	65
Aluminum	81

It is predicted that the population of the earth will plunge to about 2 billion or less from 9 billion in 30 to 100 years as a result of this over consumption of Earth's resources.

The departure of the human capitalistic system with its focus on profits in the pursuit of consumer needs and wants from Mother Nature's way of energy conservation and resource recycling has caused many unpleasant side effects that are ruining the most if not all Earths biological habitats as can be seen in the next chapter.

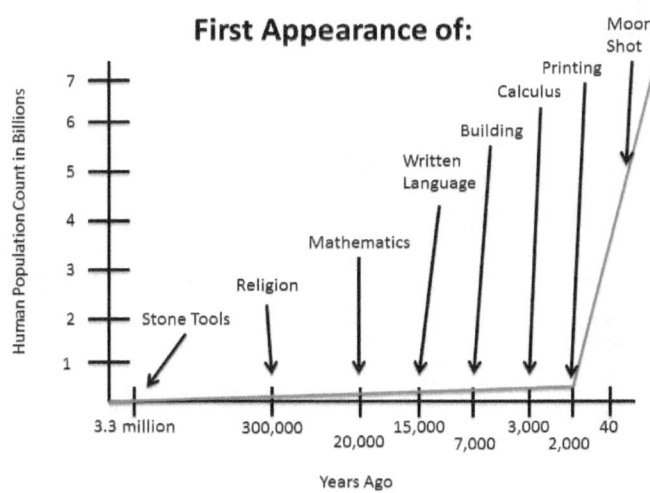

Humans, like all other creatures, behave as if they are the most important being on Earth. All the living must assume that they are at the center of the universe because any other way they would have no motivation to continue. With this new power source and motivation, humans set out to control their own destiny and proceeded to tame the world. Instead of a cyclical existence they felt comfortable with a straight line. So they built square homes, towns, and cities. They denuded the jungle and built farms with machinery that could only go in one direction and so farmland was divided into square or rectangular fields.

A natural landscape

Human landscape

Instead of being a part of the circular food chain, they chose to be at the top of a food pyramid. If a creature was thought to be useful to human existence it was domesticated or farmed. If it was not thought of as being useful to humans, it was ignored or eliminated.

Instead of being more efficient in the use of energy, humans became less so by huge infusions of fossil fuels. They built their own monetary systems based on increasing the profit margins of companies and individuals by teaching the need for material goods and services and meeting the public's demand for them (see Appendix 2). Energy was no longer a precious resource but a way to heat up economies that support material-based growing populations. Instead of being a part of the biota, they chose to be apart from it.

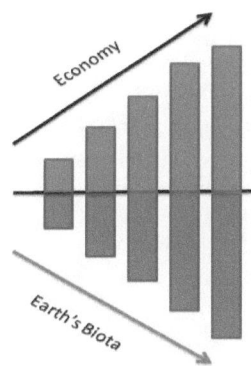

Summary

As competition to satiate the accumulation of the wealth drove the provision of more products and services for consumers an improved public health system developed, the age of death was pushed back leading to an increase in the birth rate and the population count grew exponentially. Competition also increased the quantities of goods and services for this growing population fueled by energy readily available in the form of fossil fuels. In our Zero Sum World on Earth something had to give and it was not the insatiable wants of humans but Earth's provisions of minerals and the rest of its life supporting biota.

The task in front of us now is to find out how to abate this human bloom, but first we need to discuss its dark side effects and the roadblocks to reverse this disaster.

The Dark Side of Capitalism

A person cannot help but acknowledge that economic evolution has done "well" with the human species. Humankind has successfully competed with its neighboring species leaving more offspring in subsequent generations and grew exponentially. There have been, however, indications that not all is well with this human bloom (please refer to the reading list on the prospects of the future at the end of this book). We got lost from our original physical and behavioral selves and that will be discussed in later chapters, but presently we need to talk about our effects upon the environment.

The following are but a few important examples that have popped up in the news in the past few years regarding the negative effects of the explosion of humankind's population upon the planet Earth's infrastructure; these effects destroyed multitudes of Earth's energy-efficient catalytic cycles:

Coal and Crude Oil

The use of coal and crude oil greased the skids powering the industrial revolution and every conceivable aspect of modern life from fuel for mechanical engines to plastics that seem to permeate and envelope everything we buy. This use of coal and crude oil also hailed in the haphazard research in chemistry where new and non-degradable and non-biological chemicals now poison our existence from killing precious pollinating bees to seeping into our water and food systems.

Crude oil is so pervasive because it is the lifeblood of our economies. There are three issues concerning crude oil and its products that affect the environment. The first is air pollution mainly from automobile exhaust fumes. These pollutants come from total and partial combustion of fuels. They include carbon monoxide and carbon exhaust particulates.

Animals exposed to high enough concentrations of certain air pollutants may experience:

- **Irritation of the eyes, nose, and throat**
- **Wheezing, coughing, chest tightness, and breathing difficulties**

- Worsening of existing lung and heart problems, such as asthma
- Increased risk of a heart attack

In addition, long-term exposure to air pollution can cause cancer and damage to the immune, neurological, reproductive, and respiratory systems. In extreme cases, it can even cause death.

The second issue is the spilling of crude oil and its byproducts because of human activity. Approximately 11% of these spills occur during the drilling, refining, and transportation of these products; 89% are caused by the end user in the dumping of petroleum-based products down the drain in our homes and businesses, leaky engines and mishandling of fuels in activities such as mowing the lawn or faulty storage tanks that leak into the ground.

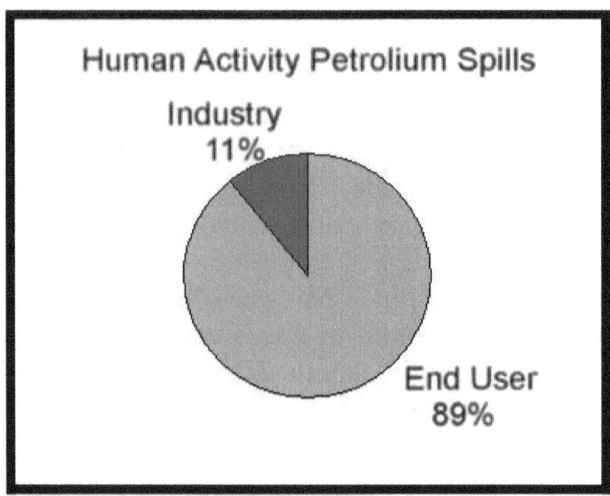

A graph showing percentages of spills due to petroleum industry and end user activities

Third, crude oil, refined petroleum products, and polycyclic aromatic hydrocarbons are ubiquitous in various environmental zones. They can bio-accumulate in food cycles where they disrupt biochemical or physiological activities of many organisms, thus causing carcinogenesis of some organs, mutagenesis in the genetic material, impairment of reproductive capacity and/or hemorrhages in exposed populations.

Tourism

Idyllic tourist destinations have been so spoiled by tourists that many authorities are being forced to close them down to prevent further damage.

For example, Koh Tachai in Thailand's Ranong Province was rated as Thailand's most beautiful island. Nevertheless, according to the Bangkok Post, the tourist-ravaged beaches of Koh Tachai will have to close indefinitely.

Tunya Netithammakul, director general of the department of national parks, wildlife, and plant conservation, said that tourism had resulted in "overcrowding and the degradation of natural resources and the environment."

The closure of the island was designed to give the land and marine environments a chance to regenerate "before the damage is beyond repair."

Mining

The environmental impact of mining includes erosion, formation of sinkholes, loss of biodiversity, and contamination of soil, groundwater, and surface water by chemicals from mining processes. In some cases, additional forest logging is done in the vicinity of mines to increase the room available for storage of the created debris and discarded soils. Besides causing environmental damage, the contamination resulting from leakage of chemicals affects the health of the local population. Mining methods have significant environmental and public health issues.

Erosion of exposed hillsides, mine dumps, tailings dams, and the resultant siltation of drainages, creeks, and rivers can significantly impact the surrounding areas, a prime example being the giant Ok Tedi Mine in Papua New Guinea. In areas of wilderness, mining may cause destruction, disturbances of ecosystems and habitats, and in areas of farming, mining may disturb or destroy productive grazing and croplands. In urbanized environments, mining may produce noise pollution, dust pollution, and visual pollution.

Mining refineries produce a large proportion of the harmful waste products, which later have to be disposed of in specific ways in order to avoid harming the environment. In addition, conventional refineries create their own toxic waste that must also be accounted for.

Governments around the globe that have some kind of "environmental" concerns and attempt to regulate mining do it to protect human safety, and in some instances, the safety of human properties not the natural fauna and flora. No concept of the cyclical nature of life on Earth is understood.

Agriculture

From the birth of agriculture, food production for humans has sequestered the natural resources of other species for human use to the point where humanity is faced with the task of becoming zookeepers of what remains of Earth's observable wildlife. The massive human population explosion is not helping but hindering the work of those who see its danger to the planet because when it comes to the last kernel of maize, the odds are that it will be given to a human and not a bird or squirrel.

Agriculture, as with any perturbation to the natural environment, has also interfered with evolution and natural selection, where naturally occurring plants and animals are replaced with domesticated ones. For instance, in the process of farming, our ancestors had a choice of picking from two varieties of wheat they collected on the hillsides of the Near East. The dominant type shattered upon the slightest touch, scattering the seeds so the species could spread and survive. The other did not scatter its seeds so easily, and therefore it was easier to handle. As a result, a higher proportion of this non-scattering variety was collected and planted than occurred in nature. With each succeeding year, a higher proportion of the non-scattering wheat was harvested and planted. Natural selection had been replaced by those who think they know best.

With agriculture, there were times when production exceeded demand, and as a result, politics was born; a decision as to who gets the surplus and what they do with it came into play, and a new world of wheeling and dealing came into being.

In summary, humans once danced to the beat of the whole Earth in unison with all life forms, as hunter-gatherer's living cycles contributed to the web of life. It is true they took but they also gave back and never took for granted what they received, and instead believed that it was a precious gift to be thankful for. The Earth was a part of them, and they were a part of it physically and spiritually.

Unfortunately, the thoughts of some humans wandered beyond the planting of seeds. One of the first wayward thoughts included domestication of animals. You may argue that it does not matter if you eat from a wild animal or a domesticated one; there is no difference pound for pound.

The breakthrough of the unconstrained Earth dancers did not stop there. The humans needed to feed their captives, and so they had to impound the grounds that others danced on. They denied those others the right to feed freely. Plants, birds, and insects were eliminated or chased away and so were the many animals that called these lands their temples.

Pesticides

Through most of history, insects have been loathed. From locusts to termites and the boll weevil they have gained a reputation as being detrimental to the interests of humankind. It was not until recent history that some insects were found to be useful and helpful to humankind.

Over 98% of sprayed insecticides and 95% of herbicides cause harm to life forms other than intended target species because they are sprayed or spread across entire agricultural fields. Runoff can carry pesticides into aquatic environments, whereas wind can carry them to other fields, grazing areas, human settlements, and undeveloped areas, potentially affecting other species. Other problems emerge from poor production, transport, and storage practices. Over time, repeated application increases pest resistance, while its effects on other species can facilitate the pest's resurgence.

Each pesticide or pesticide class comes with a specific set of environmental concerns. Such undesirable effects have led many pesticides to be banned, while regulations have limited and/or reduced the use of others. Over time, pesticides have generally become less persistent and more species-specific, reducing their environmental impact. In addition, the amounts of pesticides applied per acre have declined, in some cases by 99%. Nonetheless, the global spread of pesticide use, including the use of older or obsolete pesticides that have been banned in some jurisdictions, has increased overall.

The bee is just one example of the harm done to off-target species. It was not until its effects on pollination were discovered that value of bees' services to humankind was recognized along with that of several other insects. Today, bees and other useful insects are in danger from insecticides that are used to protect domesticated plants from other insects and fungi.

Fertilizers

Similarly to insecticides, fertilizers can run off into waterways thereby poisoning fish and other aquatic species for which fertilizers were not intended.

With the advent of the so-called Green Revolution in the second half of the 20th century—when farmers began to use technological advances to boost yields—synthetic fertilizers, pesticides, and herbicides became commonplace around the world not only on farms but also in backyard gardens and on front lawns.

These chemicals, many of which were developed in the lab and are petroleum-based, have allowed farmers and gardeners of every stripe to exercise greater control over the plants they want to grow by enriching the immediate environment and warding off "pests." But such benefits haven't come without environmental costs—namely, the wholesale pollution of most of our streams, rivers, ponds, lakes, and even coastal areas—because these synthetic chemicals run off into nearby waterways.

When the excess nutrients from all the fertilizers we use enter our waterways, they cause algae blooms sometimes big enough to make waterways impassable.

A related issue is the poisoning of aquatic life. According to the U.S. Centers for Disease Control and Prevention (CDC), Americans alone churn through 75 million pounds of pesticides each year to keep the "bugs" off their peapods and petunias. When those chemicals get into waterways, fish ingest them and become diseased. Humans who eat diseased fish can themselves become ill, completing the circle wrought by pollution.

A 2007 study of pollution in rivers around Portland, Oregon, revealed that wild salmon there are swimming around with dozens of synthetic chemicals in their system. Another recent study from Indiana showed that a variety of corn genetically engineered to produce the insecticide Bt is having toxic effects on off-target aquatic insects, including caddis flies, a major food source for fish and frogs.

Economics

Any modern economy is fueled by cheap energy scraped from or pumped out of the ground, and when the goal of this economy is to provide endless goods and services to its limitlessly self-gratifying consumers, the finite Earth must pay the price with gored landscapes, poisoned waterways and agricultural fields. This open-ended perception that the Earth is limitless in providing resources is a problem we need to face today and not tomorrow.

Extinctions

The fastest mass extinction in Earth's history is happening right now.

There is no doubt in the minds of biologists that the Earth is losing a race with time to save its precious cargo. In 1993, Harvard Biologist E.O. Wilson estimated that Earth was losing ~30,000 species per year or 3 species per hour. This means that approximately 700,000 species have gone extinct since this estimate was made. This bio-die-off crisis is snowballing, and many experts predict the sixth mass extinction on the horizon.

There is no doubt that humans are the direct cause of this habitat stress and species destruction in the modern world through such activities as transformation of the landscape (through agriculture, mining, oil exploration and fracking, construction and forestry practices), overexploitation of species (by overhunting game species), pollution, human overpopulation, and introduction of invasive species into existing habitats.

Food Chain Disruption: The most immediate impact would likely be on the food chain. For example, as sea surface temperatures continue to rise, many species of plankton are beginning to decline. If the plankton, such as diatoms and krill, were to go entirely extinct, this event would have an impact on larger creatures, such as fish and whales, who consume plankton as a major food source. In turn, if those larger marine animals have less to eat, and as their own population declines, these changes may have a domino effect throughout the food chain, ultimately reducing human food sources.

Medicine: Many medications are derived from plants, which rely on insects for pollination. As the insect population decreases, such plants may struggle to reproduce. The fewer plants there are, the harder it would be for the remaining insects to find a food source, creating a cause-and-effect loop that can hurt all the species involved. Some medications are created from the skin of amphibians, which are also under threat of extinction.

Higher Disease Risk: Some animals are resistant to certain types of diseases, e.g., the opossum's ability to ward off Lyme disease. Such buffer species help contain outbreaks from spreading to other animals and humans. As buffer species lose habitat space to urbanization or as climate-induced changes in an ecosystem proceed, the ability of such species to create the buffer declines. As they become extinct, the creatures that move in to take their place are often less adept at containing the spread of diseases or are more likely to contract such diseases, putting humans at a higher risk of diseases.

The Habitats

With the domestication of plants and animals, more and more lands are needed to support this human activity, implying a loss of a habitat for animal and plant species of no interest to humans.

For instance, in the southern United States of America, beef cattle are raised on ranges making them compete with other indigenous species that naturally occur there. This competition is aided by humankind's ability to shoot or poison those species that are undesirable to the national economy.

Soil Erosion and Degradation

Natural processes moved rocks from their resting places down hills and gullies using wind, whereas water currents caused them to collide with other rocks. All this movement resulted in the formation of soils approximately 450 million years ago, when plants started to take root after evolving from simpler organisms. Since then, soils have moved under the influence of water or wind (known as erosion) at a comparable rate of formation and deposition.

A far more recent problem is the loss of soil at a much faster rate than its formation or deposition from other areas. This phenomenon results from humankind's actions such as allowing overgrazing by cattle and other domesticated species and improper cultivation practices by means of mechanized agricultural implements. These actions leave the land fallow and unsuitable for human or natural-species use. The eroded soils are transported away and deposited in undesirable locations.

As a result of human construction activities, soils may also be degraded, which includes salinization, nutrient loss, and compaction.

Approximately 40–50% of the soil erosion on Earth is caused by humans and their irresponsible behaviors. Deforestation, improper farming techniques, intensive agriculture, and construction of roads, bridges, and dams all lead to soil erosion.

Although it is true that intensive agriculture and improper farming methods lead to soil erosion, it is equally true that soil erosion affects further agriculture. When the soil is being eroded, it is not just the soil that is taken away but the minerals and nutrients that naturally occur in the soil are also lost. Soil erosion removes the topmost layer of the soil (topsoil), which no doubt, is the most fertile and most productive part of any soil. When the top layer is removed, rills and gullies are formed in the soil.

The harmful effects of the removal of topsoil do not end there. When the top layer is removed, the soil's ability to store water and other nutrients are reduced. This situation also exposes the subsoil. This subsoil has poor physical and chemical properties making it susceptible to further erosion, and crops that were newly planted are washed off along with the soil.

This soil erosion also affects waterways. It leads to deposition of silt in the path of flowing water. The eroded soil may contain fertilizers, pesticides, and other harmful chemicals, which will decrease the quality of water in these rivers and streams.

Soil erosion compromises the soil's capacity for sustaining the growth of plants and crops; thus, agricultural yields tend to be lower. Often, the subsoil contains clay, which limits the development of the roots of plants.

Wind erosion contributes to air pollution. Soil particles carried in the air lead to dust that can contain chemicals from agricultural practices. This soil dust often leads to respiratory problems and skin infections in humans.

Ocean Pollution

This pollution occurs when harmful or potentially harmful effects result from the entry into the ocean of chemicals, particles, industrial, agricultural and residential waste, noise, or the spread of invasive organisms. Eighty percent of marine pollution comes from land. Air pollution is also a contributing factor by carrying off pesticides or dirt into the ocean. Land and air pollution have proven to be harmful to marine life and its habitats.

The pollution often comes from nonpoint sources such as agricultural runoff, wind-blown debris, and dust. Nutrient pollution, a form of water pollution, means contamination by excessive input of nutrients. It is a primary cause of eutrophication of surface waters, in which excess nutrients, usually nitrogen or phosphorus, stimulate algal growth.

Many potentially toxic chemicals adhere to tiny particles, which are then taken up by plankton and benthos animals, most of which are either deposit feeders or filter feeders. This way, the toxins are concentrated upward within ocean food chains. Many particles combine chemically in a manner highly depletive of oxygen, causing estuaries to become anoxic.

When pesticides are incorporated into the marine ecosystem, they quickly become absorbed into marine food webs. Once in the food webs, these pesticides can cause mutations, as well as diseases, which can be harmful to humans and to the entire food web.

Toxic metals can also be introduced into marine food webs. These metals can cause a change in tissue composition and biochemistry and in animal behavior and reproduction, and thereby suppress growth in marine life. Also, many animal feeds have high fish meal or fish hydrolysate content. This way, marine toxins can be transferred to land animals, and appear later in meat and dairy products.

Deforestation

It has been estimated that approximately a half of Earth's mature tropical forests—between 2.9 million and 3 million square miles of the original 5.8 million to 6.2 million square miles that until 1947 had covered the planet—have now been destroyed. Some scientists have predicted that unless significant measures such as seeking out and protecting current old-growth forests are taken on a worldwide basis, then by 2030, there will only be 10% remaining, with another 10% in a degraded condition; 80% will have been lost and with them hundreds of thousands of irreplaceable plant and animal species (along with a loss of habitat for millions of plant, animal, and insect species).

Forests are cut down for many reasons, but most of them are related to money or to people's need to provide for their families. The biggest driver of deforestation is agriculture. Farmers cut forests to gain more room for planting crops or grazing livestock. Often, many small farmers each clear a few acres to feed their families by cutting down trees and burning them in a process known as "slash and burn" agriculture.

Logging operations, which provide the world's wood and paper products, also cut countless trees each year. Loggers, some of them acting illegally, also build roads to access more and more remote forests; these actions lead to further deforestation. Forests are also cut as a result of urban sprawl.

Deforestation has many negative effects on the environment. The most dramatic impact is a loss of habitat for millions of species. Seventy percent of Earth's land animals and plants live in forests, and many cannot survive the deforestation that destroys their homes.

Deforestation also drives climate change. Forest soils are moist, but without protection from sun-blocking tree cover, they quickly dry out. Trees also help perpetuate the water cycle by returning water vapor back into the atmosphere. Without trees to fill these roles, many former forest lands can quickly become barren deserts.

Removal of trees deprives the forest of portions of its canopy, which blocks the sun's rays during the day and holds in heat at night. This disruption leads to more extreme temperature swings that can be harmful to plants and animals.

Trees also play a critical role in absorption of greenhouse gases that fuel global warming. Fewer forests mean larger amounts of greenhouse gases entering the atmosphere—and increased speed and severity of global warming.

Depletion of Global Fisheries

A report entitled *"Status and Solutions for the World's Unassessed Fisheries"* by the Sustainable Fisheries Group (SFG) has confirmed the suspicions held by many researchers that nearly 80% of the world's fisheries are in steep decline. The reasons for this decline are overfishing caused by many factors that include uninformed political pressure that tends to dominate the decision-making process, the problem of the unmanaged commons where there is no policy for management, and the use of gill nets that kills much more than the target fish. Over time, this situation can lead fisheries to collapse.

P.R. Brewbaker

Overfishing is a global problem with many serious social, economic, and environmental implications.

Environmental Effects

There is also growing evidence that the increased volume of fishing activity worldwide is having a serious effect on the health of the oceans as a whole. When commercially valuable species are overexploited, other species and habitat that share the same ecosystem are affected.

For instance, recent studies suggest that overfishing of large shark species has had a ripple effect in the shark's food chain, increasing the number of species, such as rays, that are usual prey for large sharks, resulting in declining stocks of smaller fish and shellfish favored by these species.

In addition to harvesting large amounts of fish and seafood for sale, large-scale fishing operations catch and often unintentionally kill off-target marine life, including juvenile fish, corals, and other bottom-feeding organisms, sharks, whales, sea turtles, and birds. Killing these unintended species can have significant effects on marine ecosystems.

On the basis of new information about the dynamics of marine ecosystems, more and more countries and regional fisheries management organizations (RFMOs) are adopting an ecosystem-based approach to the management of fish stocks.

Invasive Species

These are species that are transferred from one ecosystem to another and compete with native ones by taking over their habitats. An increase in global trade has increased the number of plant and animal species that are carried from one part of the globe to another legally, illegally, or accidentally. Some of these species become invasive and wreak havoc on the destination ecosystem's ecological balance.

A key factor that makes many species invasive is a lack of predators in the new environment. This factor is complex and results from thousands of years of evolution in a different place. Predators and prey often coevolve in a phenomenon called the coevolutionary

51

arms race. What this means is that as prey develop better defenses, predators in turn evolve better ways of exploiting prey. The classic example of this interplay is the cheetah and antelope. Faster antelope survive better because they can better escape cheetahs. The fastest cheetahs then survive better because they can better catch the faster antelope. Neither species ultimately gains an advantage because they continually evolve in response to one another.

Nevertheless, when a plant or animal enters a new environment, they will likely encounter predators who have not been evolving with them, which makes these predators unable to successfully exploit the prey. Defense mechanisms like venom, size, or speed that have been matched by adaptation in predators are suddenly without match in the new environment.

This advantage can allow the species to proliferate rapidly because it no longer faces any predators. Many insect or fungal species that are invasive in the United States come from regions where native trees have evolved resistance to their effects. When these species enter the US, they find trees that have no resistance and can decimate forests quickly.

Invasive species may also be able to exploit a resource that native species cannot use, which allows them to take hold in the new environment. Introduced into the Western United States, the barbed goatgrass thrives in serpentine soils, whereas native plants do not normally grow there. This exploit has given the barbed goatgrass a solid stronghold in the area. Combined with the fact that grazing animals do not like its taste, this grass has spread rapidly throughout California.

Some species also alter the environment in a manner that makes the environment more favorable for them but less suitable for natives: a phenomenon called "ecological facilitation." The yellow starthistle has been introduced into the West Coast and secretes the chemical compound 8-hydroxyquinoline from the roots. This compound harms native plants, thus allowing starthistle to increase its range as its chemicals wipe out native competitors.

P.R. Brewbaker

Hazardous and Toxic Waste

A hazardous waste is waste that poses substantial or possible threats to public health or the environment. Worldwide, the United Nations Environmental Program (UNEP) has estimated the production of more than 400 million tons of hazardous waste universally each year by industrialized countries, with approximately 4 million tons shipped across international borders. The majority of transfers from industrialized countries to developing nations have to do with disposal because of the rising cost of disposing of hazardous waste in the home country. Disposal consists of storage either above or below ground, leaving a legacy for future generations. In the case of biological toxins, they are incinerated or released into air.

Exposure to hazardous waste is harmful to humans and wildlife. It is especially dangerous for young fledgling life forms. Fetuses, whether human or animal, are in a process of rapid development. The introduction of chemicals interferes with embryonic development, resulting in malfunctioning organs and limbs.

Hazardous waste stunts plant growth, much of which is useful to humans for consumption or manufacturing. In addition, the elimination of plant life reduces the natural food supply for animals. Thus, hazardous waste exposure can potentially destroy an entire ecosystem.

The presence of hazardous waste in the environment is often the result of inadequate disposal. Improperly maintained landfills are a major issue. Although they appear to be isolated from contact with people or resources, they can still contaminate the surrounding environment. Waste in landfills sits for years or even generations. It emits gases that are both foul and toxic. It produces a liquid known as leachate, which can travel to water supplies such as rivers, lakes, and the ocean.

For example, gasoline is toxic and can cause environmental harm. It is the most popular fuel for small automobiles, and the transportation sector depends on it.

Plastics

Since their mass production began in the 1940s, plastics' wide range of unique properties has propelled them to an essential status in society. Next year, more than 300 million tons will be produced worldwide. The amount of plastics manufactured in the first 10 years of this century approaches the total produced in the entire last century, according to a report.

Evidence is mounting that the chemical building blocks that make plastics so versatile are the same components that harm people and the environment. Besides, their production and disposal contribute to an array of environmental problems. For example,

- **Chemicals added to plastics are absorbed by the human body. Some of these compounds have been found to alter hormones or to have other possible effects on human health.**
- **Plastic debris, laced with chemicals and often ingested by marine animals, can injure or poison wildlife.**
- **Floating plastic waste, which can survive for thousands of years in water, serves as mini transportation devices for invasive species, disrupting habitats.**
- **Plastic buried deep in landfills can leach harmful chemicals that spread to groundwater.**
- **Approximately 4% of world oil production is used as a feedstock to make plastics, and a similar amount is consumed as energy in the process.**

People are exposed to chemicals from plastics multiple times per day via air, dust, water, food, and use of consumer products.

For instance, phthalates serve as plasticizers in the manufacture of vinyl flooring and wall coverings, food packaging, and medical devices. Eight out of every 10 babies, and nearly all adults, have measurable levels of phthalates in their system.

In addition, bisphenol A (BPA), found in polycarbonate bottles and the linings of food and beverage cans, can leach into food and drinks. The CDC has reported that 93% of people have detectable levels of BPA in urine.

The report noted that the high exposure of premature infants in neonatal intensive care units to both BPA and phthalates is of "great concern."

Polybrominated diphenyl ethers (or PBDEs), which are flame-retardants added to polyurethane foam furniture cushions, mattresses, carpet pads, and automobile seats, are also widespread.

e-Waste

Roughly 40 million metric tons of electronic waste (e-waste) is produced globally each year, and ~13% of that waste is recycled mostly in developing countries. Approximately 9 million tons of this waste—discarded television sets, computers, cellphones, and other electronics—are produced by the European Union, according to the UNEP. This organization also notes that this estimate of waste is likely too low.

Informal recycling markets in China, India, Pakistan, Vietnam, and the Philippines handle anywhere from 50% to 80% of e-waste, often via shredding, burning, and dismantling the products in "backyards." Emissions from these recycling practices are damaging human health and the environment.

Developing countries with rapidly growing economies handle e-waste from developed countries, and from their own internal consumers. Currently, estimated 70% of e-waste handled in India is from other nations, but the UNEP estimates that between 2007 and 2020, domestic television e-waste will double, computer e-waste will increase fivefold, and cell phone waste 18-fold.

The black-market recycling practices magnify health risks. For example, primary and secondary exposure to toxic metals, such as lead, results mainly from open-air burning used to retrieve valuable components such as gold. Combustion from burning e-waste creates fine particulate matter, which is linked to pulmonary and cardiovascular diseases.

While the health implications of e-waste are difficult to isolate due to the unofficial working conditions, poverty, and poor sanitation, several studies in Guiyu, a city in southeastern China, provide relevant insights. Guiyu is known as the largest e-waste recycling site in the world, and the city's residents show substantial digestive, neurological, respiratory, and bone problems. For instance, 80% of Guiyu's children experience respiratory ailments, and are at an especially high risk of lead poisoning.

Residents of Guiyu are not the only ones at risk. Researchers such as Brett Robinson, a professor of soil and physical sciences at Lincoln University in New Zealand, warn that wind patterns in Southeast China disperse toxic particles released by open-air burning across the Pearl River Delta Region, home to 45 million people. This way, toxic chemicals from e-waste enter the "soil–crop–food pathway," one of the most significant routes of human exposure to heavy metals. These chemicals are not biodegradable: they persist in the environment for long periods, increasing exposure risk.

Global Warming

If you prepare a large capped glass flask containing ordinary air, with a green piece of filter paper inside, and expose it to sun rays, the temperature inside the flask will rise due to the various gases in the air and the color of the filter paper.

If you add pure carbon dioxide to the glass flask discussed above, the temperature inside the flask will rise more quickly. If you add pure methane, you will find that the warming effect is 21 times that of carbon dioxide. Similarly, if you add pure nitrous oxide, the warming effect is 330 times that of carbon dioxide.

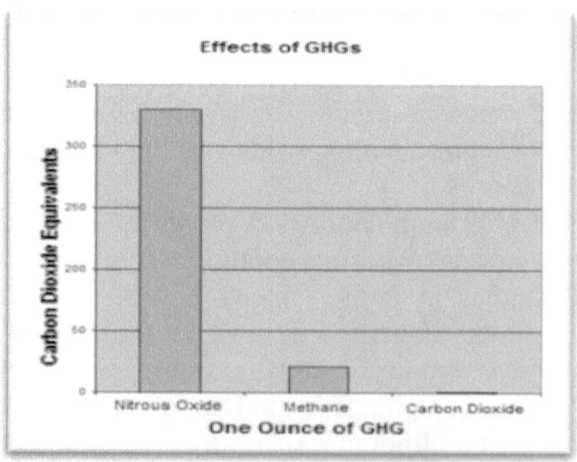

The effects of different greenhouse gases on temperature under controlled conditions in the laboratory

Today we are at the early stages of global warming and its effects, but there are signs that we are on the road to major changes in weather and ocean patterns.

Indeed, the water vapor content of the air is rising with an average increase in water vapor of .04 parts per million per year over Boulder, Colorado.

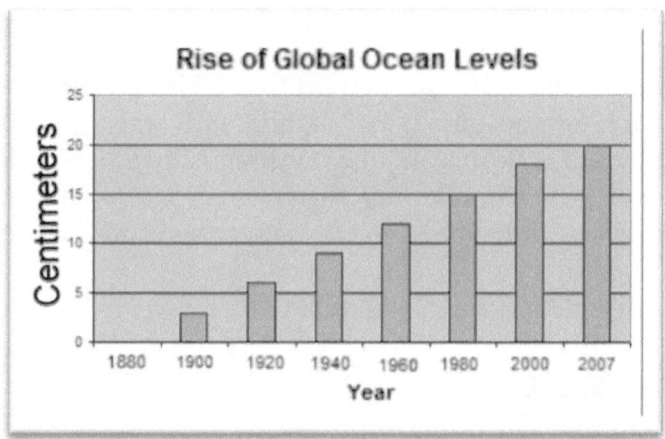

Actual sea level changes over the last one hundred years

Indeed, lower-latitude glaciers are melting away to the point where only 50% of them remain on Earth since the year 1900. There are many agricultural communities that are dependent on the annual melting of glacier waters and their replenishment.

In addition, the ocean and sea levels have been rising at the rate of an inch every 10 years, and this rise is now accelerating.

In the United States of America, forest fires are burning larger sections of land and are reaching higher up the mountains where there was once snow to slow them down. These fires are also burning hotter, and trees that depend on fires, e.g., the Ponderosa pines, are dying from the extreme heat the fires generate.

Nitrogen oxides (gases) perform an important function in atmospheric chemistry. They are emitted into the atmosphere naturally, mainly as a result of microbial activities in soils and lightning discharges. Currently and predominantly, the emissions occur as a result of human activities (such as combustion of fossil fuels, biomass burning, and the use of fertilizers). Long-term trends of their concentration in the atmosphere are not documented adequately. Nevertheless, reconstructed emission inventories suggest that large increases have occurred throughout this century. Exposure to nitrogen oxides has direct adverse effects on humans, animals, and plants. Nitrogen oxides also contribute to the global environmental problems facing our planet (i.e., excessive global warming, depletion of the ozone layer, and acid rain).

J. Hansen and colleagues report that melting of the ice sheets south and north of the globe will significantly raise ocean levels, cause severe weather storms and the hurdling of boulders into the air. These problems will reinforce our current march to global warming, which will reach a critical point in 10 years or so instead of hundreds.

Water Depletion and Contamination

Groundwater

Groundwater is a valuable resource throughout the world. Where surface water, such as lakes and rivers, is scarce or inaccessible, groundwater supplies many of the hydrologic needs of people everywhere. Sustained groundwater pumping as the result of overpopulation causes groundwater depletion and is a key issue associated with groundwater use. Many areas of the world are experiencing groundwater depletion.

Pumping groundwater faster than its recharge rate can have some negative consequences for the environment and the people who make use of the water: Lowering of the water table increases the cost of drilling and pumping of water. In addition, the reduction in water amounts in streams and lakes is partly due to groundwater depletion and lowering of the water table.

Land subsidence occurs when water is removed from the soil. This means that the soil collapses and compacts rendering it useless to agriculture or wildlife.

Deterioration of water quality in fresh groundwater supplies is caused by saltwater intrusion from oceans, seas, very deep groundwater sources, and water below oceans that is saline. (USGS Fact Sheet and USGS Circular 1186) Water quality is also affected by mining, drilling for oil, and by toxic waste dumping where heavy metals and crude oil residues seep into ground water systems.

Lake and River Water

With the increased demand for fresh water in our growing populations, we are fast approaching a point where there will not be enough water to go around. The sources for lake and river water are groundwater in the form of springs, rain, glaciers, and mountain snow. With the advent of global warming and climate change, we can see that snow and glaciers are disappearing at an alarming rate as discussed earlier. This change affects the year-round supply of groundwater, lakes, and rivers.

With global warming, the temperatures of lake and river waters are gradually, increasing causing stress to aquatic fauna and flora. This stress can lead to the death or extinction of less hardy species and perhaps to introduction of species more tolerant of warmer waters in these waterways. In turn, these phenomena affect the food supply of many terrestrial species including humans that depend on these waters for food.

Cities

Before the advent of human technology, nature had figured out how to make the population stable. This was done by having the death and birth rates more or less equal. Since then, these rules have been ignored, and now the birth rate is higher than the death rate, resulting in an explosion in the human population from 2 billion a hundred or so years ago to 7.5 billion today.

With growing populations concentrated in cities and their suburbs, communicable diseases have flourished while domestic and industrial accidents made their debut. Garbage and human fecal waste had to be rid of. Garbage is typically buried underground where it leaks toxins into the groundwater systems and into air or is taken far out to sea to be dumped. Human fecal matter is sometimes treated and returned to the Earth, but most of the time, it is dumped in rivers or into the sea.

Antibiotics and Superbugs

Antibiotic resistance is a serious and growing problem in today's health care and has become one of the eminent public health concerns of this century. In the simplest cases, microorganisms acquire resistance via mutations that make them insensitive to common antibiotics, thereby requiring the use of less common secondary antibiotics. Primary antibiotics are preferred because they have several advantages that include safety, availability, and low cost. Secondary antibiotics are broader in spectrum, pose higher risks, and may be more expensive or locally unavailable. In the case of some multiple-drug-resistant pathogens, resistance to secondary and even tertiary antibiotics has been demonstrated, as seen in pathogens *Staphylococcus aureus* and *Pseudomonas aeruginosa* (which possess strong intrinsic resistance).

The major cause of widespread antibiotic resistance is the overuse of antibiotics in agriculture and in human health care. It is becoming harder and harder to fight infections caused by these drug-resistant pathogens, and their increase in number and prevalence is outpacing society's ability to fight them.

With the use of antibiotics, we have been killing off whole biologic communities in our soils, waterways, and putting ourselves in danger.

Colonization of the world

As the accumulation of wealth became institutionalized by the technological humans and reached a fever pitch, they embarked on the great expansionary conquest of the rest of the world not realizing that they were regressing populations that had achieved a stable equilibrium with their surrounding biota. Those who had achieved a balance with nature and were dancing the cyclical dance were doomed to a life of slavery serving the technological revolution and those who proclaimed own superiority.

Genetic Engineering

This technology is interfering greatly with natural selection. For example, there is genetically modified wheat that once planted becomes sterile. The way in which this phenomenon interferes with natural selection is that natural plant pollen is used up pollenating sterile plants, and the sterile pollen of the sterile plants pollenates the natural plants hindering their natural ability to multiply.

In addition, humans are sequestering more and more land to be cultivated for these sterile plants, thereby further interfering with nature's way of doing business.

Globalization

The promotion of free trade around the world and the spread of capitalistic economic systems have increased the spread of the above-mentioned dark events around the world.

Summary

The human bloom has not been kind to the Earth. Each of the preceding examples reveals how destructive we can be on the Earth's habitats doing ordinary things like going to work in the morning or taking a vacation.

It is obvious that we have not reached the evolutionarily mature energy-efficient stage. For example, a pride of lions does not finish off a herd of gazelle and move on. The pride of lions stake out a territory and consume only a part of the gazelle herd leaving plenty of them to breed and produce naturally more food for the pride. This approach ensures efficiency in the use of the sun's energy.

The pace of technological development fueled by abundant fossil fuel energy has outpaced evolution's ability to insert one or more genes into the human genome that would mitigate the environmental problems posed by humans on Earth and in Earth's biota. This is because human genetics takes anywhere from 100,000 to 500,000 years for any viable changes to be expressed, while the changes on Earth that are leading us to an inevitable mass extinction event are only 30 to 100 years away(See Limits to Growth in reading list).

A question must be asked at this time. "Is humanity's brain, with its accelerated learning abilities, able to replace its genetically controlled underpinnings and save the Earth?" In other words, can we think our way out of our death wish and achieve an efficient and mature way to use energy?

Attempts at Turning the Tide

With the highest so-called standards of living in the societies of the developed world, not all was sugar and spice to some humans, and they started to sound the alarm. Hopes were high that something can be done about the damage humankind has caused to its precious planet.

Sustainability

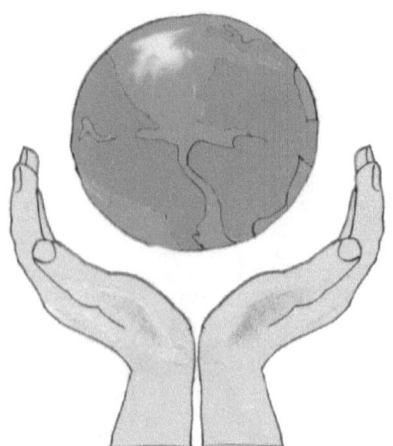

The idea of sustainability stems from the concept of sustainable development, which became a catchphrase at the World's first Earth Summit in Rio in 1992. Since then, there have been many variations and extensions on this basic definition. The following are but a few:

"Sustainability means development that meets the needs of the present without compromising the ability of future generations to meet their own needs"; Bruntland Report for the World Commission on Environment and Development (1992).

"A process of change in which the exploitation of resources, the direction of investments, the orientation of technological development and institutional change are all in harmony and enhance both current and future potential to meet human needs and aspirations"; The World Commission on Environment and Development.

"Sustainable development is a dynamic process which enables people to realize their potential and improve their quality of life in ways which simultaneously protect and enhance the earth's life support systems"; Forum for the Future.

"In essence sustainable development is about five key principles: quality of life; fairness and equity; participation and partnership; care for our environment and respect for ecological constraints – recognizing there are 'environmental limits'; and thought for the future and the precautionary principle"; from Making London Work by Forum for the Future's Sustainable Wealth London project.

"The environment must be protected... to preserve essential ecosystem functions and to provide for the wellbeing of future generations; environmental and economic policy must be integrated; the goal of policy should be an improvement in the overall quality of life, not just income growth; poverty must be ended and resources distributed more equally; and all sections of society must be involved in decision making"; the Real World Coalition 1996, a definition based on the work of the World Commission on Environment and Development.

"We cannot just add sustainable development to our current list of things to do but must learn to integrate the concepts into everything that we do"; the Dorset Education for Sustainability Network.

"A sustainable future is one in which a healthy environment, economic prosperity and social justice are pursued simultaneously to ensure the well-being and quality of life of present and future generations. Education is crucial to attaining that future"; Learning for a Sustainable Future – Teacher Centre.

"The first and perhaps most difficult problem, one that seldom gets addressed, is the time frame...Is a sustainable society one that endures for a decade, a human lifetime, or a thousand years?" (The Shaky Ground of Sustainable Development Donald Worster in Global Ecology 1993.)

"The overall goal of the UN Decade of Education for Sustainable Development (DESD) is to integrate the principles, values and practices of sustainable development into all aspects of education and learning. This educational effort will encourage changes in behavior that will create a more sustainable future in terms of environmental integrity, economic viability and a just society for present and future generations"; UN Decade of Education for Sustainable Development 2005–2014.

Many argue that sustainability has been hijacked and twisted to suit governmental and business interests that really want to continue with the way things are. For example, there is no quantifiable target of what the "sustainable future" looks like; there are no quantifiable parameters with which to measure any type of progress toward sustainability. In summary, sustainability is only a feel-good state of mind for human individuals, corporate leaders, and governments in the current state of our planet.

Conservation Biology

Conservation biology is tied closely to ecology in researching the dispersal, migration, demographics, effective population size, inbreeding depression, and minimal population viability of rare or endangered species. To better understand the restoration ecology of native plant and animal communities, a conservation biologist closely studies both their inclusive and exclusive habitats that are affected by a wide range of benign and adverse factors. Conservation biology is concerned with phenomena that affect the maintenance, loss, and restoration of biodiversity and the science of sustaining evolutionary processes that engender genetic, population, species, and ecosystem diversity. The concern stems from estimates suggesting that up to 50% of all species on the planet will disappear within the next 50 years; this situation has contributed to poverty, starvation, and will reset the course of evolution on this planet.

Conservation biologists research and educate on the trends and process of biodiversity loss, species extinctions, and the negative effect these are having on our abilities to sustain the well-being of human society. Conservation biologists work in the field and office, in government, universities, nonprofit organizations, and industry. They are funded to research, monitor, and catalog every angle of the Earth and its relation to society. The topics are diverse because this is an interdisciplinary network with professional alliances in biological and social sciences. Those dedicated to the cause and profession advocate a global response to the current biodiversity crisis based on morals, ethics, and a scientific reason. Organizations and citizens are responding to the biodiversity crisis via conservation action plans that direct research, monitoring, and education programs that deal with concerns from a local to global scale.

Environmental Protection Agencies

Environmental protection agencies or organizations have cropped up all around the world. The United States Environmental Protection Agency (EPA or sometimes USEPA) is an agency of the U.S.A. federal government which was created for the purpose of protecting human health and the environment by writing and enforcing regulations based on laws passed by Congress. The EPA was proposed by President Richard Nixon and began operation on December 2, 1970, after Nixon signed an executive order. The order establishing the EPA was ratified by committee hearings in the House and Senate. The agency is led by the Administrator, who is appointed by President and approved by Congress. The current administrator is Gina McCarthy. The EPA is not a cabinet department, but the Administrator is normally given the cabinet rank. Among EPA programs are the following:

(1)   Energy Star
In 1992, the EPA launched the Energy Star program, a voluntary program that fosters energy efficiency.

(2)   Pesticides
EPA administers the Federal Insecticide, Fungicide, and Rodenticide Act and registers all pesticides legally sold in the United States.

(3)   Environmental Impact Statement Review
The EPA is responsible for reviewing Environmental Impact Statements of other federal agencies' projects, under the National Environmental Policy Act (NEPA).

(4)   Safer Detergents Stewardship Initiative
Through the Safer Detergents Stewardship Initiative, EPA's Design for the Environment recognizes environmental leaders who voluntarily commit to the use of safer surfactants. The latter are the surfactants that break down quickly into nonpolluting compounds

and help protect aquatic life in both fresh and salt water. Nonylphenol ethoxylates, commonly referred to as NPEs, are an example of a surfactant class that does not meet the definition of a safer surfactant.

(5)   EPA Safer Choice

The EPA Safer Choice label, previously known as the Design for the Environment label, helps consumers and commercial buyers identify and select products with safer chemical ingredients, without sacrificing quality or performance. When a product has the Safer Choice label, it means that every intentionally added ingredient in the product has been evaluated by EPA scientists. Only the safest possible functional ingredients are allowed in products with the Safer Choice label.

(6)   Fuel economy

Manufacturers selling automobiles in the United States are required to provide EPA with fuel economy test results for their vehicles, and the manufacturers are not allowed to provide results from alternate sources. The fuel economy is calculated using the emissions data collected during two of the vehicle's Clean Air Act certification tests by measuring the total amount of carbon captured from exhaust fumes during the tests.

(7)   Air quality

The Air Quality Modeling Group is in the EPA's Office of Air and Radiation and provides leadership and direction on the full range of air quality models, air pollution dispersion models, and other mathematical simulation techniques used for assessing pollution control strategies and the effects of air pollution sources.

(8)   Oil pollution

The Spill Prevention, Control, and Countermeasure Rule apply to all facilities that store, handle, process, gather, transfer, store, refine, distribute, use, or consume oil or oil products. This definition includes petroleum and nonpetroleum oils as well as animal fats, oils, and greases; fish and marine mammal oils; and vegetable oils, (including oils from seeds, nuts, fruits, and kernels).

(9)   EPA WaterSense

WaterSense is an EPA program designed to encourage water efficiency in the United States through the use of a special label on consumer products. It was launched in June 2006. Products include high-efficiency toilets (HETs), bathroom sink faucets (and accessories), and irrigation equipment. WaterSense is a voluntary program, with the EPA developing specifications for water-efficient products through a public process and product testing by independent laboratories.

(10)   Drinking water

The EPA ensures safe drinking water for the public, by setting standards for more than 160,000 public water systems nationwide. The EPA oversees states, local governments, and water suppliers to enforce the standards under the Safe Drinking Water Act. The program includes regulation of injection wells in order to protect underground sources of drinking water. Selected readings of amounts of certain contaminants in drinking water, precipitation, and surface water, in addition to milk and air, are reported on EPA's Rad Net web site.

(As of this publication many of the EPA regulations are said to be rolled back erasing the progress mentioned here.)

Development of Metrics: Ecological Footprint

There have been many attempts to try to give teeth to the concept of sustainability. One of the early ones is the concept of carrying capacity of land. For example, this concept is supposed to describe how many humans can be supported on 10 acres of forest land.

Another concept is that of an ecosystem. This notion is intended to describe the relations or links between the air, soil, water, plants, and animals of a forest or a desert and in turn provide an understanding of how humans can be supported. Ecosystems are complex and hard to measure and describe fully.

Both these concepts have fallen short of giving a true picture of what it will take to support a human. With respect to the human carrying capacity of a plot of land, it was not useful because of our global economy. For example, Egypt imports 50% of its food from other countries; therefore, it is impossible to measure the carrying capacity of an acre of land in Egypt.

A recent measure reflecting the effects of humans on Earth's natural dance of rejuvenation is the ecological footprint. The latter is a standardized measure of human need for Earth's ecosystems. It measures the human demand for natural resources that are contrasted with the Earth's ability to regenerate them. It represents the amount of biologically productive land and sea areas necessary to supply the resources humans consume and to assimilate the resulting waste if a population follows a given lifestyle.

The first academic publication about ecological footprints was authored by William Rees in 1992. The ecological footprint concept and calculation method were developed as the PhD dissertation of Mathis Wackernagel, under Rees' supervision at the University of British Columbia in Vancouver, Canada, from 1990 to 1994. In 2003, Jason Venetoulis, Carl Mas, Christopher Gaudet, Dahlia Chazan, and John Talberth developed Footprint 2, which offered a series of theoretical and methodological improvements to the standard footprint approach. The four primary improvements were as follows: they included the entire surface of the Earth in bio-capacity estimates, allocated space for other (i.e., nonhuman) species, updated the basis of equivalence factors from agricultural land to net primary productivity (NPP), and refined the carbon component of the footprint based on the latest global carbon models.

Global Ecological Footprints: Currently there is no established way to measure global ecological footprints, and any attempts to describe the capacity of an ecosystem as a single number are massive simplification of thousands of key renewable resources, which are not used or replenished at the same rate. Nevertheless, there has been some convergence of metrics and standards since 2006.

That said, City Ecological Footprints measurements are being attempted. There are two types of measurements. The first evaluates

70

ecosystem displacement, which is defined as City Area minus the remaining green spaces. This is an area measure that does not include human or other biological activity. The second type is designed to quantify health of the surviving ecosystem. Specifically, it is supposed to quantify both the area and biological health of ecosystems surviving inside city areas such as nature reserves, parks, and other green spaces. City ecological footprints are being calculated and ranked with city ecological indexes.

The more detailed methods of calculating the ecological footprint are presented by Rees and Wackernagel in the book entitled *Our Ecological Footprint: Reducing Human Impact on the Earth.*

For 2007, humanity's total ecological footprint was estimated at 1.5 planet Earths; that is, humanity uses up ecological resources 1.5-fold more quickly than the Earth can renew them. We have been living off of the principal in our bank account as well as the interest. In other words, to get to the fruit, we have been cutting down its tree.

When we examine an undisturbed ecosystem, we find that each species takes only as much as can be regenerated. That means that each species has zero effect on the ecosystem and we can safely say that each species occupies an ideal zero ecological footprint.

Roadblocks

As the human bloom continues unabated, destroying every living cycle in its path, we need to ask why the human species with its intellectual prowess has not been able to produce a mature strategy and target for human existence and survival of the planet.

The Eternal Optimists

There are those of us who consider their cup perpetually full. They have every belief that faith or technology or something will always come to the rescue.

Denial

The root of denial is the lack of education and understanding of the complexities of the current situation facing us. Many who are enjoying their lifestyles are not concerned about what is going on with the environment, do not believe that it is all that serious, or have a problem believing that it is all going to end sooner rather than later.

This is especially true with seniors who consider their time on Earth limited. The younger crowds are too busy working to simply look up and see what is really going on. The optimists are always thinking that science and technology will simply solve any problems that come our way, for instance, by substituting one product for another or finding ways of cleaning up the mess we have created for ourselves.

Goals of Belief Systems

It is normal for humans to want to know the beginning, the middle, and end of things which include their story, their purpose, ideals, what happens after death, and the story of their universe. These are told in a language that the cooperative group understands, taking images from the surrounding landscape and wisdom from its leaders.

Belief systems are a reflection of what is in the hearts of humans as they try to make sense of the world. Belief systems are what guide us through our daily thoughts and activities. They give us a sense of what is right and what is wrong with respect to family, community, and strangers. They define who we are as an individual or as a nation.

Belief systems (such as the current amalgam of major religions and philosophies of life and government) rarely if ever mention the importance of maintaining or replenishing what we use of the Earth itself. One only has to look out the window and observe. Most of the Earth has been denuded of its native species, whether plant of animal, and replaced with domesticated ones. It is as if we have the right to do so and that we own this planet and can do with it whatever we please.

Goals of Governments

Governments take on the characteristics of its people or rulers and their belief systems. As with the command and control center of a cell or an animal, a modern nation has a set of laws and regulations (constitution) to be followed by all people in their pursuit of happiness. In a survey of 183 countries, most if not all defined in their constitutions in one way or another, their territories and resources, advocating the freedom of the individual to live and prosper within own territory using its resources.

Only ~50% of these countries have a constitution that mentions preservation of the environment or maintenance of the ecology within its territory. For instance, the following are statements from several constitutions in this survey:

African Nation of Namibia

1. The duty to investigate complaints concerning the overutilization of living natural resources, the irrational exploitation of non-renewable resources, the degradation and destruction of ecosystems and failure to protect the beauty and character of Namibia.

2. Maintenance of ecosystems, essential ecological processes and biological diversity of Namibia and utilization of living natural resources on a sustainable basis for the benefit of all Namibians, both present and future; in particular, the Government shall provide measures against the dumping or recycling of foreign nuclear and toxic waste on Namibian territory.

Middle Eastern Nation of Egypt

Article 44: The Nile

The state commits to protecting the Nile River, maintaining Egypt's historic rights thereto, rationalizing and maximizing its benefits, not wasting its water or polluting it. The state commits to protecting its mineral water, to adopting methods appropriate to achieve water safety, and to supporting scientific research in this field.

Every citizen has the right to enjoy the Nile River. It is prohibited to encroach upon it or to harm the river environment. The state guarantees to remove encroachments thereon. The foregoing is regulated by law.

Article 45: Seas, beaches, lakes, waterways, mineral water and natural reserves

The state commits to protecting its seas, beaches, lakes, waterways, mineral water, and natural reserves.

It is prohibited to encroach upon, pollute, or use them in a manner that contradicts their nature. Every citizen has the right to enjoy them as regulated by law. The state also commits to the protection and development of green space in urban areas; the protection of plants, livestock, and fisheries; the protection of endangered species; and the prevention of cruelty to animals. All the foregoing takes place as regulated by law.

Article 46: Environment
Every individual has the right to live in a healthy, sound and balanced environment. Its protection is a national duty. The state is committed to taking the necessary measures to preserve it, avoid harming it, rationally use its natural resources to ensure that sustainable development is achieved, and guarantee the rights of future generations thereto (22 Draft dated 2 December 2013 of the Constitution of the Arab Republic of Egypt).

The European Nation of Finland
Section 20 – Responsibility for the environment
Nature and its biodiversity, the environment and national heritage are the responsibility of everyone. The public authorities shall endeavor to guarantee for everyone the right to a healthy environment and for everyone the possibility to influence the decisions that concern their own living environment.

Constitution Summary

On average, if you add up all the constitutional texts of the 186 countries surveyed, the statements involving the preservation of the environment and its ecology represent pitiful 0.2% of the sentence count. Again, we see that an overwhelming majority of the laws on Earth today do not reflect the reality: a finite and cyclical nature of its biota. It is true that some nations have organizations that are actively trying to protect the environment like environmental protection agencies, but these have mostly been knee jerk reaction agencies trying to rectify massive damage that has already been done lacking specific goals to actually maintain a healthy ecosystem or eliminate humanity's damaging ecological footprint.

Goals of Corporations and Companies

As with belief systems and governments, businesses have a figurehead who orchestrates the business plan and mission statement or goals that have been written down for all involved to follow. The following is a small sample of mission statements of leading businesses around the world taken from their websites. Again, concern for well-being of the Earth, its environments, and a reduction in humanity's ecological footprint are lacking.

Bruegger

Bruegger bakes authentic New York-style bagels fresh from the oven all day. With the coffee freshly brewing and a full selection of breakfast and lunch sandwiches, wraps and Paninis, specialty soups and salads, our team creates a memorable experience that makes people say, "Let's go to Bruegger's!"

Nike

"To lead in corporate citizenship through proactive programs that reflect caring for the world family of Nike, our teammates, our consumers, and those who provide services to Nike."

Unilever

Unilever USA, the parent company of Lipton Tea, outlines its mission as working to create a better future every day helping people feel and look better and get more out of life with brands and services that are both good for them as well as others, inspiring people to take small actions that make big differences in the world, and developing new ways to double the business, as well as reduce the environmental impact of the business.

International Business Machines (IBM)

At IBM, we strive to lead in the invention, development and manufacture of the industry's most advanced information technologies, including computer systems, software, storage systems and   microelectronics.

We translate these advanced technologies into value for our customers through our professional solutions, services and consulting businesses worldwide.

Hertz

"Our mission is to be the most customer focused, cost efficient vehicle and equipment rental/leasing company in every market we serve. We will strengthen our leading worldwide positions through a shared value culture of employee and partner involvement by making strategic investments in our brand, people and products. The focus of everything we do will be continuously improving shareholder value."

Boeing

Leadership

We will be a world-class leader in every aspect of our business—in developing our team leadership skills at every level; in our management performance; in the way we design, build and support our products; and in our financial results.

Integrity

We will always take the high road by practicing the highest ethical standards and by honoring our commitments. We will take personal responsibility for our actions and treat everyone fairly and with trust and respect.

Quality

We will strive for continuous quality improvement in all that we do so that we will rank among the world's premier industrial firms in customer, employee and community satisfaction.

Customer satisfaction

Satisfied customers are essential to our success. We will achieve total customer satisfaction by understanding what the customer wants and delivering it flawlessly.

People working together

We recognize that our strength and our competitive advantage is—and always will be—people. We will continually learn, and share ideas and knowledge. We will encourage cooperative efforts at every level and across all activities in our company.

A diverse and involved team

We value the skills, strengths and perspectives of our diverse team. We will foster a participatory workplace that enables people to get involved in making decisions about their work that advance our common business objectives.

Good corporate citizenship

We will provide a safe workplace and protect the environment. We will promote the health and well-being of Boeing people and their families. We will work with our communities by volunteering and financially supporting education and other worthy causes.

Enhancing shareholder value

Our business must produce a profit, and we must generate superior returns on the assets entrusted to us by our shareholders. We will ensure our success by satisfying our customers and increasing shareholder value.

Core competencies

Detailed customer knowledge and focus

We will seek to understand, anticipate and be responsive to our customers' needs.

Large-scale systems integration

We will continuously develop, advance and protect the technical excellence that allows us to integrate effectively the systems we design and produce.

Lean enterprise

Our entire enterprise will be a lean operation, characterized by the efficient use of assets, high inventory turns, excellent supplier management, short cycle times, high quality and low transaction costs.

Strategies

Run healthy core businesses.

Leverage strengths into new products and services.

Open new frontiers.

People working together as a global enterprise for aerospace leadership.

Mission Statement Summary

Out of the six above-mentioned mission statements for leading companies on Earth, only one mentions the environment, and even in that one case, it is only to reduce and not eliminate the company's impact. Again, we find a lack of urgency in matters of the ecological stability of the planet. Perhaps this is because the bottom line of every business is its profitability compared to its competitors. Profitability is where the company takes more than what it gives out. This profit is essential for growing the company, and the growth is another goal that every for-profit organization tries to achieve. Perhaps there is still a belief that the resources on Earth are vast and endless.

Resistance to Change

As is often the case in the pursuit of sustainability or conservation, if the reason for a change is ambiguous, whether it is about costs, equipment, or jobs, it can trigger negative reactions among people.

Another reason people resist change is when they have not been consulted or involved or when the change is forced upon them. Change also carries a threat to established patterns of working and social relations among people, or it has not been communicated well to the people involved.

The third reason change is a threat is when the benefits and rewards for making the change are not seen as worth the trouble involved. In addition, when the change threatens jobs, power, or status in an organization, it is doomed to fail.

When change affects people in power more than it does the regular Joe, these people put up the highest resistance.

Colossus Economies

Sketch of a P&H Mine Pro Shovel

Please visit Appendix 2, which discusses the birth of economic thought, where a question is posed: "Where do you think this dynamic endless loop of supply and demand has led us?" The answer is that it has led us to what can be described by one phrase, "*Economies of Scale.*" In other words and generally speaking, it has been discovered that the greater the scale of the manufacturer or producer, the lower is the cost per unit output.

An old example may clarify the point. Adam Smith published a book in 1776 on economics entitled *The Wealth of Nations* describing pin making. He explained that one pin maker with little experience can make one to ten pins a day. Then, he described a group of 18 people all involved in the making of pins where they divided the process into 18 steps, where one person performed a single step in the process, the result being that this group was able to manufacture 48,000 pins a day.

As a result, the cost of making a pin was far lower when the company of pin makers was larger and they adopted the methods described above that are also known as *The Division of Labor.* A modern example is large jumbo jets that carry 400 to 600 passengers at a time. To an airline, the cost per seat or mile is much lower in comparison with an aircraft that carries 100 passengers, and the reduced cost is passed on to the consumer.

When several companies are competing to provide the lowest cost item to the public who are growing in numbers and who want cheaper goods or services with equal or better quality, these companies tend to go for economies of scale, and generally speaking, building or purchasing larger and larger equipment, factories, or products with the end results discussed in the section on *The Assault* and its undesirable outcomes.

The sad part of economies of scale is that we are locked into them, and it is hard to go back. Reversing economies of scale is not like downsizing a company to reduce cost but like raising the cost of living considerably. It just does not make sense to the average person.

It is a good feeling when a person is able to make a cheaper widget or purchase one. It is natural after all. Nonetheless, this natural feeling is carried forward from a time when higher energy efficiency was the norm. The value of fossil fuels dictated by monetary systems is interfering with our ability to devise ways to truly sustain or conserve a lifestyle suitable for the Earth.

The Dichotomy

On the one hand, we have the companies and corporations that are competing to make less expensive widgets and the consumers trying to find reasonably priced ones. As economies of scale come into play, we find companies moving to locations where they can build mega factories and where the public knows them only by brand name, with human populations increasingly moving to larger cities.

This separation of the producers from the consumers creates a vacuum of information requiring an additional layer of government watchdogs to make sure that producers are behaving properly in developed countries. This state of affairs becomes more complex when the producers go global and in most instances face no watchdogs at all.

This vacuum of information makes it exceedingly difficult for any decisions on the well-being of the planet to be carried out. On the one hand, we have the developed country's consumers enjoying less expensive widgets and the third world countries only glad to fulfill this need by accepting the mega factory's promise of a better life for its people.

It is true that the United Nations has succeeded (after many years) in reaching a consensus on carbon emission controls that have yet to be instated and will take many more years to implement, but carbon emissions are only a fraction of the problem as we saw in the section on *The Assault*. This situation highlights the problem in the approach to solving the catastrophic environmental problems we face. There is no consensus on the root cause of our environments problems, and instead we are trying to tackle one symptom at a time out of a large multitude.

Cataclysm by Capitalism

Tolerance

I lived in the city of Springfield, Oregon, back in the 1970s. It is a small city in comparison to Portland, but that was an advantage because it was at the far south end of the Willamette Valley where the coastal and Cascade Mountains met. It was a beautiful scene in timber country.

Unfortunately, when the wind was coming from the east, the stench of rotten eggs would overcome me. Yet no one complained. When I asked why, I was told by my grandfather that it was the smell of money. You see, the paper plant then was the largest employer.

Humans have a great ability to accept an inconvenience if in the end they benefit. It is only when this nuisance is coupled with the knowledge of personal danger that something is done. And so it was with the smog that filled the air in the 1960s and 70s when large cities started to be affected by the exhaust of autos and power plants. It wasn't until the dangers of lung diseases from such polluted air became known that the danger signs went up causing the introduction of emission controls and transplanting of coal or oil electric generator plants away from large cities.

When we think of the benefits of using buses for transportation within cities, that is a good thing. But when we think of competition in making cheaper buses leading to economies of scale, with its assault on the environment, that is a bad thing. We get confused by the need to hop on the do-good bandwagon only to find out that it also does us and the environment harm.

We are therefore unable to think of a conservation target in favor of the environment because it will eventually bite us where it hurts: in the pocket book. Some of us try to think of targets only to find out they eventually will not work because of our dependence on the colossal economy. For example, I have read books on how to "leave civilization" and live off of one to three acres of land by farming it one way or another. Unfortunately, at the end of the day, we need to go back to our homes (that take forest land to grow) with all their modern-day conveniences (steel pots and pans that require mining of iron and processing in mega factories).

Environmental Politics

It is a sorry state when the speaker of the United States of America's House of Representatives says that environmental scientists are using pseudo-science to make their predictions of the calamities facing our nation and the globe. Our scientists are the brains of our country, and all they are doing is measuring and reporting on what they see. They don't have an ax to grind.

It is therefore exceedingly difficult for our nation to make any kind of informed decisions on the environment with such attitudes of people wielding so much power.

Apathy

Some have realized that the damage done to the biosphere can't be reversed, and consequently they give up and voice the opinion that any suggestions to change our ways of life are fruitless, will not happen, or are irrational. They claim that there have been several catastrophic events on Earth in the past 4 billion years, so why should we worry about this one. After all, those living today will not be around when it happens.

Intentions and Birth Control

There have been attempts to gain control of the human population in the 1960s and 1970s. This was mostly done in third world counties that were thought to have an exploding population. The attempts incorporated birth control options that included sterilization.

This effort backfired. Many third world countries objected to mandatory birth control especially sterilization on the grounds that it discriminated between rich and poor nations. Poor nations felt that rich nations were imposing their will upon the poor.

That is not to say that sterilization was not attempted in the United States of America. It suffered a similar fate.

Summary

The roadblocks that are keeping humans from formulating a global goal or target for our societies that would take us all out of our current tragic situation are many, and so the human bloom continues unabated undermining the cyclical nature of life on Earth. Whenever a voice is raised on behalf of the Earth regarding what action needs to be taken to formulate an all-inclusive target for all to follow, there is a dozen voices from individuals, companies, and governments, who support our current lifestyles, cry foul, , or "it is pseudo-science", or "it can't be done" or are silent.

There is also ignorance of what the basic problems are. For example, very few realize that the nature of life on Earth consists of energy efficient cycles that can be studied, measured, and quantified. Without this knowledge, a unified global or local societal target can never be formulated.

Thus, a unified target to save the biota (of which we are an integral part) on Earth has never been achieved. The next part of this publication explains a fair and equitable societal target to aim for that takes into consideration the needs of humans as well as the needs of the biota on Earth.

Target

Mother Earth has been dancing in the light of the Sun, is billions of years old and through it all has learned a thing or two, by building self-correcting behaviors into the biota to maintain the dance of life giving pleasures to the participants and hardships to those who veer away.

After the great assault of the humans and the spreading of their ways of assuming that every setback is a challenge to their notions of grandeur instead of a lesson in how to become wise, they stopped listening to the subtle tempo of the biotic music that was being played and tried to overcome it. It did not matter what catastrophe swept through their overcrowded communities: no locust swarms or crop failures or famine or plague stood in their way. They had to overcome nature as if it was an enemy and not their friend that brought them news of the right path to a peaceful and rich life of play, fun, rhythm, and song.

Humans had expansionistic technologies on their side boosting their egos with thoughts of splendor and magnificence thinking that these technologies contained a kernel of wisdom on which the world ran; that nature was so-called primitive and not of human form, to be despised and trampled upon. How wrong they were and are!

Despite nature's repeated attempts to sound an alarm, humans continued to poison the streams of advice flowing though the lands now filled with toxic hopes of reaching asteroids for the minerals that will become exhausted here on Earth. So, now human veins are carrying poisoned blood, and human thoughts are concentrated on where else to dump the outcomes of their actions so that nothing can be seen. Technology has robbed them of a future free of foolishness.

Humans once had friends with which they played and danced to the music of blissful existence, but since they chose to cease brotherly relations with the rest of life forms, they have no one to pull them back into the fold of mutual love and respect. They stand alone in their quests with no one to help them out when they fall. They have cut all ties with their family and have no one to lean on when the inevitable comes.

The humans had to toil with that which can't be toiled with. It will not be until the last individual is born that they will realize that they should have listened to the Mother, the Father, and the rest of their kin.

Return to Nature

Be advised that this section and those that follow are not for the faint-hearted. They contain language and ideas that are contrary to most of current sensibilities. They present facts about our current and future lifestyles. They discuss a unified global lifestyle to save the biota on Earth and much of what our future is to be like if we are to survive but does not propose how we get there. That task is up to you.

Zero Ecological Footprint

This means that in a biological cycle what is taken locally is returned in full locally.

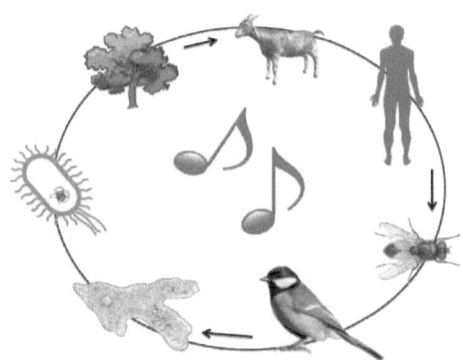

For example, if we take the diagram above, the starting point is the bush, which converts sunlight into bioenergy. The bush also takes up minerals from the Earth. It is then partially eaten by the goat that is partially eaten by the human, and so on. The feces of the human are then partially consumed by the fly's offspring followed by the bird eating the flies and so on around the cycle. The following table shows a simulation of the cycling of energy and minerals. It implies that the bush has 100 units of energy (of which 50% are consumed by the goat) and 10 units of minerals (50% of which are also consumed by the goat). In addition, only 80% of the consumed energy and minerals are metabolized, and the rest are defecated.

Consumption			Used		Defecated		Totals	
Species	Consumed	Metabolized	Energy	Mineral	Energy	Mineral	Energy	Mineral
Goat	50%	80%	40.00	4.00	10.00	1.00	50.00	5.00
Human	50%	80%	16.00	1.60	4.00	0.40	20.00	2.00
Fly	50%	80%	1.60	0.16	0.40	0.04	2.00	0.20
Bird	50%	80%	0.64	0.06	0.16	0.02	0.80	0.08
Amoeba	50%	80%	0.06	0.01	0.02	0.00	0.08	0.01
Bacteria	50%	80%	0.03	0.00	0.01	0.00	0.03	0.00
Bio-degradable			21.67	2.17	5.42	0.54	27.09	2.71

The balance sheet for the above cycle

The last column in the balance sheet above represents the total amounts consumed (metabolized and defecated) and not used (biodegradable material that becomes a part of the ground humus). Of course, the balance sheet above is a simplification of the processes in the cycle. For example, it doesn't show catabolism, but you can build your own table that includes it.

The take-away insight from the diagramed cycle and its balance sheet above is that energy and minerals are conserved. Of course, you must realize that energy metabolized can serve many functions not demonstrated here. It can be used for movement, warmth, and perspiration as well as building muscle or hair. In addition, readers must realize that all energy and minerals are conserved <u>locally</u>.

The Target Human Population Count

The truth of the matter is that humans evolved into a diverse biota, and their genetic makeup had to adapt. Once one thinks along the lines of humans' being just one aspect of Earth's diversity and that all life is interlocked in the cyclical dance of existence, it becomes clear that we all depend on each other and that breaking of even one dancing cycle will ultimately unravel the whole web of existence.

Therefore, the question of the target should consider humans not as the top of the food chain but as an equal and integral part of life on planet Earth. Notions like "humans have superior intelligence" have continuously crept into the literature as an expression of our naturally blind drive to acquire more resources than our perceived competitors, enemies, or anything that stands in our way. It must be emphasized that humans are a link no better or worse than any other link, be it a clump of mud or another species in the web of life. As for human intelligence, it has been subservient to the expansionistic life stage present in all species (be they amoebas, birds, or dinosaurs): to steal, hoard, and defeat the enemy to provide for themselves and their young.

This expansionistic life stage worked well for humans at their early developmental stage when the Earth seemed vast and endless, and the opportunistic spirit was alive and well. Today, it is neither vast nor endless, and we are fast approaching critical limits of the current expression of our expansionistic life stage in a finite world. We must find a way to move on with our development to the mature energy-efficient stage and at the same time to reverse our paths from those that created these critical limits and to work to eliminate them. This means that we must reverse course and build up the principal in our bank accounts. We must share the interest with all other species. Instead of cutting down the tree to get to the fruit, we should wait for the fruit to ripen and perhaps fall to the ground before we eat it.

Our aim should be to have no impact on the Earth. Our aim is to have zero ecological footprint like that of the lioness we covered previously. As it turns out, the ideal human population count should be between 10 and 50 million individuals (see Appendix 4). This estimate is based on available land and sunlight. Upcoming chapters will describe the lifestyle and other conditions needed to support this population count at zero ecological footprint.

Human Niche

We need to get in touch with our original selves out from the shadows of our artificial modern lives. We will never know for sure what the original human niche was like because that bridge has been crossed and was washed away maybe 10,000 years ago. We can't stop wondering, though, what it was like to fit the environment like a hand fits a finely tailor-made glove because it was in that niche that we were the most efficient. It was in that niche that we enjoyed an inborn freedom as simple as plucking a wild apple from its tree and not worrying where the next one will come from. That niche is where natural zero ecological footprint was the norm.

You might ask, "Has the human condition changed with respect to its original niche?" The answer is yes and no. As far as our genetics, it probably has not changed that much because of the length of time it takes for a viable genetic trait to appear on its own; something like 100,000 years. But as far as the physical environment is concerned, it has changed, as discussed in the section on *Darkness*. In addition, our behaviors have changed because today our lives are wrapped up in large societies that are different from those of the hunter gatherer people's small groups, as you will see.

There are clues to what it must have been like. We can look at our closest relatives who are alive today or who perished not long ago and imagine. From there, we can look at how different we are from them and what that must have meant and therefore see the difference between our and their niches. We introduced the hunter gatherer groups at the beginning of this book and will now look at the chimpanzee. It is hoped that the reader will discover semblances of his or her physical, behavioral characteristics and the environment for which they are better suited.

The Chimpanzee: Physical Characteristics

(1) Chimpanzee Natural Habitat
The habitat of the chimpanzee is one that is diverse. The rainforests are the most common areas where they are found. Nonetheless, they also live in the savannahs and in mountain forest areas. They are adaptable to a variety of environments. This is why their habitat continues to be one that is vast and diversified.

In a rainforest, they are able to find plenty of trees for hiding and for sleeping. It also makes it easier for them to gain access to fruits, plants, leaves, and seeds as well as shaded areas. In savannahs, chimpanzees are in open areas that have scattered woodlands.

A lower number of chimpanzees are found in mountainous regions because of the climate. It is also possible that some moved there in search of food, once their habitat had been taken from them by humans or other species.

Chimpanzees do need water, and therefore they will be found in areas where there is enough food and a good supply of water. This supply can come from a stream of major rivers.

A chimpanzee troop usually has a home range. The size of this area can vary from ~5 to 400 km^2. It often depends on the regenerative properties of the area and the number of chimpanzees in the troop. When the home range is large, the group will often split up and forage into different areas of the range.

The emigrational pattern through the home range is the one that is complex. It ensures that chimpanzees can get to certain areas where seasonal fruits grow just in time. The movement prevents them from depleting any single area of the food sources that are available there. In fact, their methods of eating often ensure that new vegetation can grow back for future use.

(2) The Fossil Record

New research demonstrates that hominins (early human species) in what is today northern Africa lived equally well in a relatively warm and dry climate 3.4 million years ago and in a much cooler climate with significantly more rainfall and forest growth slightly later. *Australopithecus afarensis* adapted to these dramatic environmental changes without the benefit of an enlarged brain, which aided later hominins in adapting to their environments.

These findings contribute to an ongoing debate about whether hominins of the Pliocene era preferred settings that were open and arid or wooded and moist, or whether they could adapt well to diverse environments. A lack of data on changes in past ecosystems to compare with hominin fossil data has hampered the inquiry.

Researchers analyzed fossil pollen located in stratified rock formations near Hadar, Ethiopia. Using these samples, the researchers identified three persistent plant communities: steppe and tropical and temperate forests containing water-conserving plants. A fourth plant community, forests containing plants that grow in cooler and wetter climates, appears and disappears in the pollen record. The presence of this fourth community corresponds to climate records of cooler and wetter periods.

These early humans had the ability to adapt to environmental changes. They could live in arid grasslands and forested surroundings as well.

Other researchers reconstructed likely responses of human ancestors to the climate of the past 5 million years using genetic modeling techniques. When results were mapped against the timeline of human evolution, the researchers found that key adaptive events coincided with periods of high variability in recorded temperatures.

P.R. Brewbaker

The study confirmed that a major human adaptive radiation—a pattern whereby the number of coexisting species increases rapidly before crashing again to near previous levels—coincided with an extended period of climatic fluctuation. After the onset of high climatic variability ~2.7 million years ago, a number of new human species appeared in the fossil record, with most disappearing by 1.5 million years ago. The first stone tools appear at ~3.2 million years ago and doubtless assisted some of these species in responding to the rapidly changing climatic conditions.

By 1.5 million years ago, there was a single human ancestor left: *Homo erectus*. The key to the survival of *H. erectus* appears to be its behavioral flexibility: it is the most geographically widespread species of the period, and endured for over 1.5 million years. While other species may have specialized in environments that subsequently disappeared—causing their extinction—*H. erectus* appears to have been a generalist, able to deal with many climatic and environmental conditions.

Variability selection suggests that evolution, when faced with rapid climatic fluctuation, responded to the range of habitats encountered rather than to each individual habitat in turn; the timeline of variability selection suggests that *H. erectus* could be a product of exactly this process.

Linking climatic fluctuation to the evolutionary process has implications for the current global climate change debate. What we see in many areas of the world today is in fact an increased annual range of temperatures and conditions; this means in particular that third world human populations, many living in what are already marginal environments, will face ever more difficult situations. The current pattern of human-induced climate change is unlike anything we have seen before and is disproportionately affecting areas whose inhabitants do not have the technology required to deal with this change.

Physical Evidence Summary

Judging by the physical evidence, it appears that our early ancestors had several niches to deal with because of the constantly changing environment. They became jacks-of-many-trades. This transition required changes to their morphology such as a larger brain for adaptation, bipedal mobility to be more efficient in long treks, and an increase of meat and fat in their diet for more energy as evident from their shorter gut.

Another way to put it is that at an early stage, we became generalists coping with a parade of several niches that we had to adapt to. As a result, the forest people stayed in the forest, while others roamed the deserts like today's nomads in the Sahara but retained the ability to interbreed.

Behavioral Characteristics

There are three major factors influencing our behaviors. First, there are our emotions, which were borne within us over thousands if not millions of years in response to the environments we found ourselves in and preserved in our genetics.

Second, there are the factors of physical actions taken to express our emotions also stored away in our genetics and evolved over thousands if not millions of years in our struggle to adapt to the parade of niches we found ourselves in.

Third, there is a belief system that supports the verbal or physical expression of these emotions colored by the local environmental conditions.

For example, when an individual takes action to discriminate against another this action evolved to protect the individual and the tribe or clan from a foe. That behavior is not recognized today as legitimate but was legitimate then and is a clue to our true niche.

You see when we feel hate, love, or indifference; we must realize that these feelings evolved in us over long periods of time for the purpose of participation in nature's orchestra and performing the cyclical dance. Emotions, actions, and belief systems are nature's answer to the song that defines us and our niche.

We must peel away the layers of inhibitions cast upon us by those who blindly saw and built mega nations that invented colossal machines to rape the Earth's treasures.

Looking for the original human behaviors adapted to their niche of zero ecological footprint is much harder than looking for the physical characteristics because there is very little in the way of artifacts. What is available to us from our past are the behaviors of our closest relatives (chimpanzees), the results of our tool making, some pictures on the walls of caves, the behaviors of our cohorts hunter-gatherers, and our own "modern" human behaviors to give us some clues.

Chimpanzees

The chimpanzee is humans' closest living relative. These two species look alike in many ways, both in body and behavior. But for a clear understanding of how closely they are related, scientists compared their DNA, an essential molecule that's the instruction manual for building each species. Humans and chimps share a surprising 98.8% of their DNA. How can we be so similar and yet so different?

Humans and chimps descended from a single ancestor species that lived 6 or 7 million years ago. As humans and chimps gradually evolved from a common ancestor, their DNA, passed from generation to generation, changed too. In fact, many of these DNA changes led to differences between human and chimp appearance and behavior.

Chimpanzee groups in different areas share different cultures much like human cultures. Tool use is a good example. Chimpanzees in one area use long twigs and alter them for better termite fishing. In a different forest chimpanzees are more often seen nut-cracking with rocks on flat surfaces.

Chimpanzees develop different cultural practices depending on their environment, and transmit these practices as learned behavior. Chimpanzees have exhibited as many as 39 learned behaviors, including feeding, mating, grooming, and tool use.

But what kinds of behaviors do most sub-species and groups of chimpanzees share? Group structuring, communication, and hunting practices are often common behaviors from one chimpanzee group to another. In addition, these behaviors aren't perfectly identical and have adaptive variations.

Chimpanzee behavior is so complex because a chimp's mental capacity is so greatly developed. Through research, many mental traits that were once considered unique to humans have been demonstrated by chimpanzees, such as reasoned thought, abstraction, generalization, symbolic representation, and a concept of self.

One of the most important discoveries was that chimpanzees make and use tools. For example, chimps pick up small twigs, strip off the leaves, and use them as tools to fish for termites in the ground, which they then sweep into their mouths as a snack.

In general, chimpanzees use objects—stems, twigs, branches, leaves, and rocks—in nine different ways to accomplish tasks associated with feeding, drinking, cleaning themselves, investigating out-of-reach objects, and as weapons.

Chimps also make sponges by chewing leaves and then dipping them into puddles of water, which they can then drink. They also use sticks and rocks to smash fruits or hard shells. Tools, however, are not used only for feeding. Adult males, for example, can enhance charging displays by hurling sticks, branches, or rocks to intimidate others.

These behaviors, passed from one generation to the next through observational learning, can be regarded as examples of chimp culture. Chimpanzees are highly social beings, just like humans. Social interactions are essential in a chimpanzee's development, learning, and overall well-being.

Chimpanzees live in social groups called communities or unit troops. At one wildlife preserve, the number of individuals has ranged between 40 and 60. Communities may be smaller or larger in other areas.

Chimpanzees' social structure can be categorized as "fusion–fission." This means that they travel around in small subgroups (of up to 10 chimps) whose membership is always changing as individuals wander off on their own or join other groups. At times, many of a community's members come together in large excited gatherings, usually when fruit is available in one part of the range, or when a sexually popular female comes into a period of female sexual receptivity.

Individuals may switch groups on occasion, but close, supportive, affectionate bonds also develop between family members and other individuals within a community and can last a lifetime. Chimpanzee family bonds are strong, especially mother–daughter bonds. Mothers and their dependent young up to age seven or so are always together. Some individuals travel together more often than others, such as siblings and pairs of male friends. Contact is maintained between members of the scattered groups by means of a distance call: the pant hoot.

Females disperse from the natal group once they are mature and spend most of their time alone, with dependent offspring. Males usually remain in birth groups, cooperate in defense of the community range, and spend long periods in proximity to other males. Males will sometimes form coalitions to support one another during conflicts with other groups.

Within a chimp community, a male hierarchy, ordered more or less in linear fashion, establishes social standing, with one male at the top or "alpha" position. All adult males dominate all females although females have their own hierarchy, albeit much less straightforward.

Age is a deciding factor in male dominance hierarchies: the alpha male is usually between the age of 20 and 26. Other factors that determine dominance and social status are physical fitness, aggressiveness, fighting skills, the ability to form coalitions, intelligence and other characteristics. Status is either maintained or changed through communication and social interactions, such as physical competition and grooming.

The males of a community regularly patrol their territories, and if they encounter individuals of a neighboring community, they may attack with extreme brutality. The only individuals who can move freely between communities are adolescent females who have not yet given birth. They may transfer to a new community permanently or, having become pregnant, move back to their own birth group.

There are several mating patterns seen in chimps. Some females in the period of sexual receptivity are more attractive than others. A popular female may be accompanied by many or all the adult males of her community, with adolescents and juveniles tagging along. Or, the dominant male of the group may show possessive behavior toward her, trying to prevent other males from mating with her. A third mating pattern is when a male persuades a female to accompany him to a peripheral part of the community range. If he can keep her there until the time of ovulation, he has a good chance of siring her child. Even low-ranking males can become fathers in this way, if they have the skill to lead a female away during the fertile period of her reproductive cycle.

Given that there is a hierarchical system in chimp societies, most disputes within a community can be solved by threats rather than actual attacks. They use gestures and postures to indicate a threat, such as tipping the head, making hitting gestures, flapping hands in the air, swaying branches, throwing objects, and charging toward another. These gestures are often combined with vocalizations.

Chimpanzees, however, are capable of physical violence. In one situation, a 4-year territory war between two groups of chimps ended with one group's killing all members of the other group. This was the first recorded account of nonhuman primate warfare.

Infanticide also occurs within chimp communities. Male chimps sometimes kill infant chimpanzees, for a variety of hypothetical reasons, but this behavior is most commonly thought to induce the female who he is mating with to wean his offspring sooner and to ensure that the offspring is his.

Chimpanzees communicate with a wide range of verbal and nonverbal means, such as calls, postures, and gestures. Chimpanzees use verbal communication, such as alarm calls, mating calls, and greeting vocalizations. So far, researchers have identified more than 30 calls, which can sometimes be heard up to 2 miles away.

Food calls (a mixture of grunts, barks, and pant hoots) alert other chimpanzees to the whereabouts of food sources. A special intensity of excited calls of this type indicates that there has been a successful kill after a hunt.

A loud, long "wraaaa" call is made when a chimpanzee comes across something unusual or dangerous. When young chimpanzees play, they emit breathy laughter. And soft grunts uttered by foraging or resting chimpanzees probably serve to maintain communication within the group.

Each individual has his or her own distinctive pant hoot, so that the caller can be identified with precision. Postures, gestures, and facial expressions communicate many messages and emotions within a group.

When greeting a dominant individual after an absence or in response to an aggressive gesture, nervous subordinates may approach with submissive signals (crouching, presenting the rump, holding the hand out) accompanied by pant grunts or squeaks. In response, the dominant individual is likely to make gestures of reassurance, such as touching, kissing, or embracing the subordinate.

Friendly physical contact is crucial in maintaining good relationships among chimpanzees. For this reason, social grooming is probably the most important social behavior, serving to sustain or improve friendships within the community and to calm nervous or tense individuals. The grin of fear seen in frightened chimpanzees may be similar to the nervous smiles given by humans when tense or in stressful situations.

When angry, chimpanzees may stand upright, swagger, wave their arms, throw branches or rocks: all with bristling hair and screaming or with lips bunched in ferocious scowls. Male chimpanzees proclaim their dominance with spectacular charging displays during which they slap their hands, stamp their feet, drag branches as they run, or hurl rocks to perhaps intimidate a higher-ranking individual without having to fight.

Chimps are omnivores and eat not only fruits, nuts, seeds, blossoms, and leaves but many kinds of insects and occasionally the meat of medium-sized mammals. Chimpanzees, like humans, have such diverse tastes that they are able to live in a wide variety of habitats, unlike gorillas and orangutans, which have more specialized diets in the wild.

Male chimpanzees occasionally hunt cooperatively, to kill and eat other mammals, such as bush pigs, monkeys, and antelope. For instance, a group of chimpanzees attacked, killed, and ate a red colobus monkey that had climbed high into a tree. The hunters covered all available escape routes while one adolescent male crept up after the prey and captured it, whereupon the other males instantly rushed up and seized parts of the carcass.

Successful hunters typically share some portion of their kill with other group members in response to a variety of begging behaviors. Most of the captured animal is eaten, including the brain. Meat is a favored food item among chimpanzees, but makes up less than 2% of their overall diet.

99

An interesting observation has also been made where chimps deliberately eat medicinal plants like *Aspilia* leaves and minerals to relieve stomach pains or reduce the number of internal parasites.

Chimps are also capable of walking upright on two legs, like humans. Unlike humans, they usually use this mode of locomotion only if they need to travel while carrying objects in their hands.

Summary

Our original zero ecological footprint niche was one of hunter-gatherer nomadic small groups consisting of several families that foraged on fruits, berries, mushrooms, some grains, flowers, and game. Hunter-gatherers move on when these food sources reached a stage of reduction but were still able to recover.

These small groups had territories that were defended fiercely because the territories were needed for survival as small groups and did not represent an object of wealth or status. The territory protected their most simple of natural cyclical lifestyles.

Tools were made of local perishable materials consisting of tree branches, roots, animal skins, animal bones, and stones. The process of making these tools also involved perishable and local materials to guarantee their return to nature. The same goes for the construction of dwellings and shelters.

They had a symbolic language and therefore thought, as demonstrated by the art found in caves and by contemporary hunter-gatherer groups. Today's hunter-gatherer groups are healthy and free of modern human diseases such as obesity, hypertension, diabetes, and communicable diseases.

They had music and song in their lives that told stories of the universe, the hunt as well as those of love and appreciation. They had rituals for preserving the group and the environment. You may think of this as their religious demonstration.

The context of the hunter-gatherer lifestyle within its environment is the one that best describes the natural human niche that is in balance with nature supporting the zero ecological footprint way of life through the cyclical character of life on Earth.

The Proposal

Now that the music for our dancing cycles has stopped and was replaced by the clanking noise of gargantuan machines, it will take some effort to return to how it once was. It is also said that "You can never go home again." The best we can hope for is a new way of life perhaps a little different from what it was 40,000 years ago and certainly radically different from what it is today.

How we get there from here will be discussed in the next chapter. For now we need to cover the rules under which we need to live and thrive followed by the benefits reaped thereof.

The first thing is our attitude. We must become wise to the fact and consider ourselves a part and not apart from the living cycles, the web, and the biota on Earth. Notions like "humans are at the top of the food chain" and "humans have superior intelligence" have continuously crept into everyday talk and the literature as an expression of our naturally blind drive to acquire more.

It must be emphasized that humans are not at the top of the food chain but they are a link no better or worse than any other link, be it a clump of mud or another species in the web of life. As for human intelligence, it has been subservient to the opportunistic drive borne in all species, be they amoebas, birds, or dinosaurs. Now is the time to forgo this opportunistic lifestyle and develop a mature attitude.

Second, each one of us must identify their living cycle. It is the living cycle of the omnivore or those who consume plant and animal matter for sustenance and leave behind only the waste of plant and animal matter for our brethren who use it for their sustenance.

Third, we must not ingest or consume anything that did not originate from our local living cycle. That means we must not ingest anything processed, manufactured, or imported because this kind of nutrition only returns us into the endless loop of self-destruction. We must thank our sources of food, Earth, and the sun because they are the providers.

Fourth, we must not leave behind anything that can't be consumed by those who are next in line in our local living cycle. It shouldn't be processed or manufactured.

Fifth, we must not use implements that do not come from our local living cycle. We can use anything that occurs naturally such as local stones, tree branches, leaves, and the like.

Cyclical Lifestyle Benefits:

The following are the benefits gained by the human condition following the five rules stated above. They don't cover the obvious benefits of the planet Earth and the rest of its biota.

- We will hunt and gather around where we live and so we do not have to leave home base or be away from family and friends.
- Hunting and gathering takes about 2 to 4 hours a day and all the food is fresh. With a short work day, we have abundant leisure time to be with family and friends.
- Food is always fresh and better than our "Organic"-labeled foods: without hormones, pesticides, or fertilizers.
- The elderly spend their days not alone but as a part of the group they have always lived with.
- Hunters and gatherers are always physically fit, and there is no problem of excess body weight.
- Because hunter-gatherers require large areas as their territories, we won't suffer from the sedentary-lifestyle diseases of civilization.
- Because we won't have any possessions, we will have no worries about anyone stealing.
- Men get to exercise their instincts for hunting and fishing.
- Men and women get to exercise all their instincts for living.
- Get rid of the stress of modern life.
- Forgo death by technology.
- Forgo living in an untested, ever changing, and artificial world.

- Enjoy personal satisfaction that each one of us will be living an Earth-sustainable lifestyle that will last for thousands if not millions of generations to come.

- Enjoy personal satisfaction that each one of us will be living a lifestyle that is a gift of life to thousands if not millions of generations to come.

- Experience personal satisfaction that each one of us will have available to them wholesome and healthy foods whose sources will last for thousands if not millions of generations to come.

- Live a more energy-efficient lifestyle.

Only when we live this simple life, will we sway to the music and join the rest of the biota in the dance once again. So, praise the sun and mother Earth because we are their precious children one and all.

Path

It would help to define what the population size target is for this chapter. We have stated in a previous chapter that the most energy-efficient and zero ecological footprint population size needs to be from 10 to 50 million human individuals scattered across the Earth adopting the hunter-gatherer lifestyle or something similar. Given that the current population is approximately 7.5 billion, the task in front of us is extremely daunting to say the least.

We have two major options in front of us at this time. The first is to do nothing. There is evidence that nature takes care of its own problems, and therefore this approach might be a viable way to go and we will explore this option presently from a logical point of view using the data we have at hand. The second option of doing something that will be covered later.

The United Nations did a simulation of birth and death rates in the human population, which indicated that the human population bloom will peak at 10 billion, at which point births will equal deaths. That is only ~35% more than what the human population count is right now. Some may think that this is a small increase that may be supported. At that point, however, humanity will be consuming twice what the Earth can produce. This means that the rate of consuming the infrastructure of Earth will double.

Other simulations indicate that this situation will lead to degraded living conditions. For example, health care and basic food needs will not be accessible to a growing number of humans, the death rate will continue to grow, and the population will decrease until it gradually reaches ~2 billion humans in approximately 50 years.

Along the way from a population level of 7.5 to 10 then down to 2 billion humans, resources will be scarce; the probability of social unrest will increase, leading to degradation of the social fabric, an increase in anarchy, and the specter of nuclear wars. These scenarios will lead to an uninhabitable Earth in the form of a decimated nonhuman species and a poisoned physical environment as discussed earlier.

The second option is to do something to avert the destruction of our Earth and its inhabitants as they stand today. To be effective we need to include all humans in the solution we need to educate everyone. All people need to know the tragic consequences of doing nothing. They also need to know how they can help to find a solution to (and prevent) the upcoming disaster.

In the face of the upcoming shortages in economically extractable metals and minerals from the Earth, it has been suggested that humanity mine these resourced from other planets or asteroids. Unfortunately, this approach adds to the energy consumption of each human and thereby decreases their efficiency and goes against the Earthly bio-fuel-efficiency trend.

Some advocate sending a group of astronauts to a distant planet to start a human colony. That idea is natural. It represents our opportunistic spirit much like the slime mold's fruiting body using the wind or other animals to carry its spores to distant territories richer in resources. Unfortunately, planets that are assumed to support Earthlike living conditions are light years away, and it will take more than 100 years of research to make such a journey possible, meanwhile this effort will be interrupted by famine and unrest hampering any hopes of funding such an expensive endeavor.

As discussed, the use of fossil fuels to run our industrial economies cannot come about without the support of the technology creation sector. They figured out how to use energy to make an aspirin tablet to sending men to the moon on a fuel-guzzling rocket. They increased the fossil fuel consumption from zero to 14 billion calories per person per year making humans by far the least efficient consumer of energy of all the living on planet Earth. They should have the opportunity to redeem themselves and create technologies that will reduce fossil fuel consumption back to zero again.

And indeed, they have been trying. There has been a lot of talk of a carbon tax that imposes a monetary penalty on carbon emissions from burning of fossil fuels. The way it works is that a government agency taxes fossil fuel companies (that pump this fuel or dig it out of the ground). The amount of tax is dependent on the amount of carbon emissions the fuel will eventually produce when used. That agency then pays companies for the "scrubbing" the carbon before or after it escapes into the air.

Unfortunately, the carbon tax does not decrease the production of fossil fuels, and in some cases, increases it. This situation adds to the human bloom and its side effects of consuming the infrastructure of the Earth and decreases the overall efficiency of humans.

What is needed is a method of gradually <u>reducing</u> fossil fuel use to zero. One way to do that is to reduce carbon emissions by 70 pounds per person per year every year for 30 years. This approach will result in two achievements. The first is that in the year 2031, there will be zero fossil fuel use. The second is that possibly the planet will be saved and able to recover.

Possible Outcome if We Aim for Zero Fossil Fuel Use in 30 years

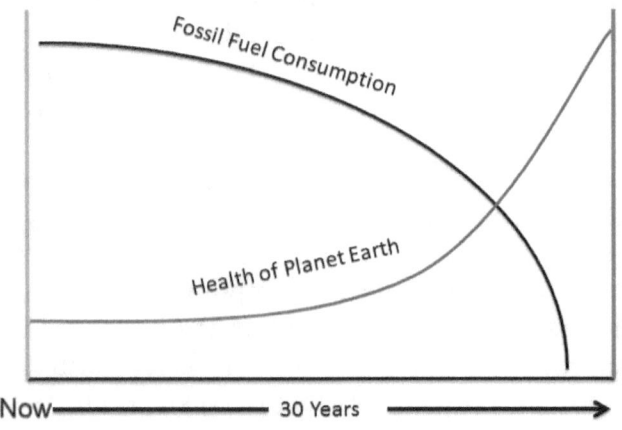

Another opportunity is the altruistic act of voluntary zero births through mass contraception except for a select group of humans who will act as our seed for generations to come much like the slime mold's fruiting body. The difficult task here is the selection process. It is our hope that all so called "primitive" tribes be exempt from this effort. They have survived at zero ecological footprint the longest, are best equipped physically, socially and have the right belief systems to be the most suitable seed.

No single approach from the opportunities presented above will work by itself. The best opportunities are those of education, voluntary birth control and the efforts to eliminate fossil fuel use are the best. The path to these goals is difficult to say the least causing social disruption and mass population relocation since the present world economies are largely dependent on these fuels. But we have a duty to insure a favorable outcome for the Earth's biota including our own sooner rather than later.

Action Plans

Education

Education is the oracle for the future. Without it we are sailing on a ship with no compass. This education must be one of integrity and not based on alternative facts. It must be rest upon the truth and not political aspirations. It must emanate from those who dedicate their lives in the pursuit of the scientific method and not the consensus of the masses.

It is true the scientific community is struggling in the wake of the revival of faith based political activists who want to protect the status quoi and change things to the way they were in the nostalgic past. Unfortunately that is like having the kidneys dictating the judgements of the brain. What should be happening is that decisions be made based on our scientifically evaluated information. That is the way nature intended it to be.

All species on Earth have an information system based on knowledge. Single celled species have a genetically based information system that was honed by experience over millions of years. More complex species require a central nervous system for gathering information and making decisions in a shorter time frame. The more complex species, like humans, have a brain that receives external and internal evidence to make timely choices on how they should respond; the more accurate the incoming data the better the resultant actions.

To the human species, the scientific method is currently the only approach to understand the truth of nature as opposed to consensus of the masses that only yields mediocrity. Governments should consist of meek men of science, 45 years old serving two years each as conscripts with a modest salary. Only then can the human species move in the required directions of zero ecological footprint, restoring the cyclical nature of life on Earth and become efficient energy users once again. For example, in the United States of America, the constitution should be changed so that all House

Representatives, Senators and the President should be replaced by such scientists.

The tasks before these scientist rulers are to educate the public in the upcoming dark 30 years, prepare them for almost zero births and zero fossil fuel use. Of course the reason why all these changes need to occur has to be loud and clear. The public must be involved and engaged on a global scale.

Almost Zero Human Population Birth Rate Plan
The aim is to reduce the birth rate to zero for those populations that used fossil fuels. It is proposed that every birth is to be taxed at one year's salary of the father.

Zero Fossil Fuel Use Plan
The aim is to reduce fossil fuel use to zero in 30 years. It is proposed that each consecutive fossil fuel production year be reduced by 23.9% of the previous year's total production. This will achieve zero fossil fuel production in 30 years. Any fossil fuel company producing an excess of this reduced amount will be taxed 23.9% of its total production for that year.

The above are my proposed action plans. What are yours?

Now is the time we should start being efficient catalysts and rejoin the rest of the biota claiming our equitable allotment of the Sun's blessing. Now is the time we start achieving a zero ecological footprint and nestle back in with our brothers and sisters for a long term future. Now is the time to wholeheartedly rejoin the symphony of life on Earth and dance in unbroken cycles.

Appendix 1

Evolution

How is evolution achieved? How does lowering the amount of energy in a chemical reaction attained? The answer is in the availability of a catalyst. In most cases where a catalyst is required, reactions occur faster because they require less activation energy. The following six parts explain how this occurs.

Part one: The application of energy

Chemical reactions are happening naturally all the time. For example, as the temperatures increases in any particular environment chemical reactions occur more frequently. If chemicals require 10 units of energy (this is called the activation energy) to react and the available energy in the environment is 8 units the chemical reaction will not occur. If the energy available in the environment increases to 11 units then the chemicals will react.

Part two: The Catalysts

Catalysis is the increase in the rate of a chemical reaction due to the participation of an additional substance called a catalyst. In most cases where a catalyst is required, reactions occur faster because they require less activation energy. In the example above, if the activation energy for a chemical reaction is 10 units and the available energy in the environment is 8 units then the reaction will not occur. In the event that a catalyst is present in this environment which lowers the activation energy for the reaction to 7 units the reaction will proceed. Furthermore, catalysts are not consumed in catalyzed reactions therefore they can continue to act repeatedly. Often only tiny amounts are required in principle.

Part three: The Enzymes

Enzymes are a subset of catalysts. They accelerate, or catalyze, bio-chemical reactions by lowering the activation energy required for the reaction. The molecules at the beginning of the process upon which enzymes act are called substrates that are converted into different molecules, called products. Almost all metabolic processes in a living cell need enzymes.

Part four: Efficiency
It has been demonstrated that catalysts lower the activation energy of a reaction and we can say that a chemical reaction that takes less energy to convert substrate to product is also more efficient in the use of energy.

Part five: Effects of Efficiency
There is a relationship between catalysts and evolution. Say you have two catalysts in an environment. Call them C1 and C2 that catalyze products C1 and C2 respectively. C1 lowers the activation energy by 2 units causing reactions that produce 10 C1s while C2 lowers the activation energy by 4 units resulting in reactions that produce 1000 C2s. After 4 replications C2 will outnumber C1 by a factor of 16.

Part six: Self Replication
Furthermore, catalyst C2 was the daughter of C1 which was changed to be more efficient by an environmental agent.

To summarize, evolution is a process of self-replication of catalysts where a change or changes occur to the offspring that lead to more efficient catalysis.

Appendix 2

Birth of Economic Thought

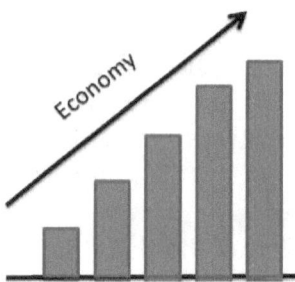

According to Hamid S. Hosseini, the power of supply and demand was understood to some extent by several early Muslim scholars, such as 14th-century A.D. Mamluk scholar Ibn Taymiyyah, who wrote: "If desire for goods increases while its availability decreases, its price rises. On the other hand, if availability of the good increases and the desire for it decreases, the price comes down."

John Locke's 1691 work, *Some Considerations on the Consequences of the Lowering of Interest and the Raising of the Value of Money* includes an early and clear description of supply and demand and their relation. In this description, demand is rent where the price of any commodity rises or falls by the proportion of the number of buyers and sellers and that which regulates the price of goods is nothing else but their quantity in proportion to their rent.

The phrase "supply and demand" was first used by James Denham-Stuart in his Inquiry into the *Principles of Political Economy*, published in 1767. Adam Smith used the phrase in his 1776 book, *The Wealth of Nations*, and David Ricardo titled one chapter of his 1817 work, *Principles of Political Economy and Taxation*, "On the Influence of Demand and Supply on Price."

In *The Wealth of Nations*, Smith generally assumed that the supply price was fixed but that its value would decrease as its scarcity increased, in effect what was later called the law of demand. Ricardo, in *Principles of Political Economy and Taxation*, more rigorously laid down the idea of the assumptions that were used to build his ideas of supply and demand. Antoine Augustin Cournot first developed a mathematical model of supply and demand in his 1838 research into the *Mathematical Principles of Wealth*.

In his 1870 essay, *On the Graphical Representation of Supply and Demand*, Fleming Jenkin introduced a diagrammatic method into the English economic literature by publishing the first drawing of supply and demand curves, including comparative statistics from a shift of supply or demand and application to the labor market. The model was further developed and popularized by Alfred Marshall in the 1890 textbook *Principles of Economics*.

The Supply and Demand Chart

Within the context of supply and demand is the need to increase the bottom line of those supplying the products and in turn the consumers who are demanding cheaper units. To the manufacturer or producer, this means a decrease in the cost of producing a unit. To the consumer, it means a decrease in the cost of purchasing a unit.

Appendix 3

The modern capitalist system has its origin in the "crisis of the fourteenth century", a conflict between the land-owning aristocracy and the agricultural producers, or serfs. Manorial arrangements inhibited the development of capitalism in a number of ways. Serfs had obligations to produce for lords and therefore had no interest in technological innovation; they also had no interest in cooperating with one another since they produced to sustain their own families. The lords who owned the land relied on force to guarantee that they were provided with sufficient food. Because lords were not producing to sell on the market, there was no competitive pressure for them to innovate. Finally, because lords expanded their power and wealth through military means, they spent their wealth on military equipment or on conspicuous consumption that helped foster alliances with other lords; they had no incentive to invest in developing new productive technologies.

The demographic crisis of the fourteenth century upset this arrangement. This crisis had several causes: agricultural productivity reached its technological limitations and stopped growing, bad weather led to the Great Famine of 1315–1317, and the Black Death of 1348–1350 led to a population crash. These factors led to a decline in agricultural production. In response, feudal lords sought to expand agricultural production by expanding their domains through warfare; they therefore demanded more tribute from their serfs to pay for military expenses. In England, many serfs rebelled. Some moved to towns, some purchased land, and some entered into favorable contracts to rent lands from lords who needed to repopulate their estates.

The collapse of the manorial system in England enlarged the class of tenant farmers with more freedom to market their goods and thus more incentive to invest in new technologies. Lords who did not want to rely on renters could buy out or evict tenant farmers, but then had to hire free labor to work their estates, giving them an incentive to invest in two very different kinds of commodity owners: those who had money, the means of production, and subsistence, who were eager to valorize the sum of value they had appropriated by buying the labor power of others; and the free workers, who sold their own labor power and, therefore, labor. The free workers were "free workers" in the double sense that they neither formed part of the means of production nor did they own the means of production that transformed land and even money into what we now call "capital". Marx labeled this period the "pre-history of capitalism".

In effect, feudalism began to lay down some of the foundations necessary for the development of mercantilism, a precursor of capitalism. Feudalism was mostly confined to Europe and lasted from the medieval period through the sixteenth century. Feudal manors were almost entirely self-sufficient, and therefore limited the role of the market. This stifled any incipient tendency towards capitalism. However, the relatively sudden emergence of new technologies and discoveries, particularly in agriculture and exploration, facilitated the growth of capitalism. The most important development at the end of feudalism was the emergence of what Robert Degan calls "the dichotomy between wage earners and capitalist merchants". The competitive nature meant there are always winners and losers, and this became clear as feudalism evolved into mercantilism, an economic system characterized by the private or corporate ownership of capital goods, investments determined by private decisions, and by prices, production, and the distribution of goods determined mainly by competition in a free market.

Rise of towns

The transition from feudalism to early forms of capitalism happened in periods differing by country. According to Cambridge political philosopher and historian Quentin Skinner, the towns of northern Italy were the first urbanized parts of Europe, dating to the twelfth century. German bishop Otto of Freising recorded the growth of town life in Germany, the loyalty of landed nobility to town authorities, and the emergence of republicanism and the belief in civic liberty.

Agrarian capitalism and enclosure

England in the sixteenth century was already a centralized state, in which much of the feudal order of Medieval Europe had been swept away. This centralization was strengthened by a good system of roads and a disproportionately large capital city, London. The capital acted as a central market hub for the entire country, creating a very large internal market for goods, instead of the fragmented feudal holdings that prevailed in most parts of the Continent. The economic foundations of the agricultural system were also beginning to diverge substantially; the manorial system had broken down by this time, and land began to be concentrated in the hands of fewer landlords with increasingly large estates. Instead of a serf-based system of labor, workers were being employed as part of a broader and expanding money economy. The system put pressure on both the landlords and the tenants to increase the productivity of the agriculture to make profit; the weakened coercive power of the aristocracy to extract peasant surpluses encouraged them to try out better methods, and the tenants also had incentive to improve their methods, in order to flourish in an increasingly competitive labor market. Terms of rent for the land were becoming subject to economic market forces rather than the previous stagnant system of custom and feudal obligation.

An important aspect of this process was the enclosure of the common land held in the open field system where peasants had traditional rights, such as mowing meadows for hay and grazing livestock. Once enclosed, these uses of the land became restricted to the owner, and it ceased to be land for commons. The process of enclosure began to be a widespread feature of the English agricultural landscape during the 16th century. By the 19th century, unenclosed commons had become largely restricted to rough pasture in mountainous areas and to relatively small parts of the lowlands.

Marxist and neo-Marxist historians argue that rich landowners used their control of state processes to appropriate public land for their private benefit. This created a landless working class that provided the labor required in the new industries developing in the north of England. For example: "In agriculture the years between 1760 and 1820 are the years of wholesale enclosure in which, in village after village, common rights are lost". "Enclosure (when all the sophistications are allowed for) was a plain enough case of class robbery".

Other scholars argue that the better-off members of the European peasantry encouraged and participated actively in enclosure, seeking to end the perpetual poverty of subsistence farming. "We should be careful not to ascribe to developments that were the consequence of a much broader and more complex process of historical change." "The impact of eighteenth and nineteenth century enclosure has been grossly exaggerated...."

Merchant capitalism and mercantilism

The earliest recorded activity of long-distance profit-seeking merchants can be traced back to the Old Assyrian merchants active in the 2nd millennium BCE. The Roman Empire developed more advanced forms of commerce, and similarly widespread networks existed in Islamic nations, but capitalism took shape in Europe in the late Middle Ages and Renaissance. However, while trade has existed since early in human history, it was not capitalism.

An early emergence of commerce occurred on monastic estates in Italy and France and in the independent city republics of Italy during the late Middle Ages. Innovations in banking, insurance, accountancy, and various production and commercial practices linked closely to a 'spirit' of frugality, reinvestment, and city life, promoted attitudes which sociologists have tended to associate only with northern Europe, Protestantism and a much later age. The city republics maintained their political independence from Empire and Church, traded with North Africa, the Middle East and Asia, and introduced Eastern practices. They were also considerably different from the absolutist monarchies of Spain and France, and were strongly attached to civic liberty.

Emergence

Modern capitalism, however, only fully emerged in the early modern period between the 16th and 18th centuries, with the establishment of mercantilism or merchant capitalism.

Early evidence for mercantile practices appears in early modern Venice, Genoa, and Pisa over the Mediterranean trade in bullion. The region of mercantilism's real birth, however, was the Atlantic Ocean.

England began a large-scale and integrative approach to mercantilism during the Elizabethan Era. An early statement on national balance of trade appeared in Discourse of the Common Weal of this Realm of England, 1549: "We must always take heed that we buy no more from strangers than we sell them, for so should we impoverish ourselves and enrich them." The period featured various but often disjointed efforts by the court of Queen Elizabeth to develop a naval and merchant fleet capable of challenging the Spanish stranglehold on trade and of expanding the growth of bullion at home. Queen Elizabeth promoted the Trade and Navigation Acts in Parliament and issued orders to her navy for the protection and promotion of English shipping.

These efforts organized national resources sufficiently in the defense of England against the far larger and more powerful Spanish Empire, and in turn paved the foundation for establishing a global empire in the 19th century. The authors noted most for establishing the English mercantilist system include Gerard de Malynes and Thomas Mun, who first articulated the Elizabethan System. The latter's England's Treasure by Foreign Trade, or the Balance of our Foreign Trade is The Rule of Our Treasure gave a systematic and coherent explanation of the concept of balance of trade. It was written in the 1620s and published in 1664. Mercantile doctrines were further developed by Josiah Child. Numerous French authors helped to cement French policy around mercantilism in the 17th century. French mercantilism was best articulated by Jean-Baptiste Colbert (in office, 1665–1683), although his policies were greatly liberated under Napoleon.

Doctrines

Under mercantilism, European merchants, backed by state controls, subsidies, and monopolies, made most of their profits from the buying and selling of goods. In the words of Francis Bacon, the purpose of mercantilism was "the opening and well-balancing of trade; the cherishing of manufacturers; the banishing of idleness; the repressing of waste and excess by sumptuary laws; the improvement and husbanding of the soil; the regulation of prices..." Similar practices of economic regimentation had begun earlier in the medieval towns. However, under mercantilism, given the contemporaneous rise of absolutism, the state superseded the local guilds as the regulator of the economy.

The Anglo-Dutch Wars were fought between the English and the Dutch for control over the seas and trade routes.

Among the major tenets of mercantilist theory was Bullionism, a doctrine stressing the importance of accumulating precious metals. Mercantilists argued that a state should export more goods than it imported so that foreigners would have to pay the difference in precious metals. Mercantilists asserted that only raw materials that could not be extracted at home should be imported; and promoted government subsidies, such as the granting of monopolies and protective tariffs, were necessary to encourage home production of manufactured goods.

Proponents of mercantilism emphasized state power and overseas conquest as the principal aim of economic policy. If a state could not supply its own raw materials, according to the mercantilists, it should acquire colonies from which they could be extracted. Colonies constituted not only sources of supply for raw materials but also markets for finished products. Because it was not in the interests of the state to allow competition, to help the mercantilists, colonies should be prevented from engaging in manufacturing and trading with foreign powers.

Mercantilism was a system of trade for profit, although commodities were still largely produced by non-capitalist production methods. Noting the various pre-capitalist features of mercantilism, Karl Polanyi argued that "mercantilism, with all its tendency toward commercialization, never attacked the safeguards which protected the two basic elements of production – labor and land – from becoming the elements of commerce"; thus with mercantilism regulation was more akin to feudalism than capitalism. According to Polanyi, "not until 1834 was a competitive labor market established in England, hence industrial capitalism as a social system cannot be said to have existed before that date."

Chartered trading companies

The Muscovy Company was the first major chartered joint stock English trading company established in 1555 with a monopoly on trade between England and Muscovy. It was an offshoot of the earlier Company of Merchant Adventurers to New Lands, founded in 1551 by Richard Chancellor, Sebastian Cabot and Sir Hugh Willoughby to locate the Northeast Passage to China to allow trade. This was the precursor to a type of business that would soon flourish in England, the Dutch Republic and elsewhere.

The British East India Company (1600) and the Dutch East India Company (1602) launched an era of large state chartered trading companies. These companies were characterized by their monopoly on trade, granted by letters patent provided by the state. Recognized as chartered joint-stock companies by the state, these companies enjoyed power, ranging from lawmaking, military, and treaty-making privileges. Characterized by its colonial and expansionary powers by states, powerful nation-states sought to accumulate precious metals, and military conflicts arose. During this era, merchants, who had previously traded on their own, invested capital in the East India Companies and other colonies, seeking a return on investment.

Industrial capitalism

Mercantilism declined in Great Britain in the mid-18th century, when a new group of economic theorists, led by Adam Smith, challenged fundamental mercantilist doctrines as the belief that the amount of the world's wealth remained constant and that a state could only increase its wealth at the expense of another state. However, in more undeveloped economies, such as Prussia and Russia, with their much younger manufacturing bases, mercantilism continued to find favor after other states had turned to newer doctrines.

The mid-18th century gave rise to industrial capitalism, made possible by the accumulation of vast amounts of capital under the merchant phase of capitalism and its investment in machinery. Industrial capitalism, which Marx dated from the last third of the 18th century, marked the development of the factory system of manufacturing, characterized by a complex division of labor between and within work process and the routinization of work tasks; and finally established the global domination of the capitalist mode of production.

During the resulting Industrial Revolution, the industrialist replaced the merchant as a dominant actor in the capitalist system and affected the decline of the traditional handicraft skills of artisans, guilds, and journeymen. Also during this period, capitalism marked the transformation of relations between the British landowning gentry and peasants, giving rise to the production of cash crops for the market rather than for subsistence on a feudal manor. The surplus generated by the rise of commercial agriculture encouraged increased mechanization of agriculture.

Industrial Revolution

The productivity gains of capitalist production began a sustained and unprecedented increase at the turn of the 19th century, in a process commonly referred to as the Industrial Revolution. Starting in about 1760 in England, there was a steady transition to new manufacturing processes in a variety of industries, including going from hand production methods to machine production, new chemical manufacturing and iron production processes, improved efficiency of water power, the increasing use of steam power and the development of machine tools. It also included the change from wood and other bio-fuels to coal.

In textile manufacturing, mechanized cotton spinning powered by steam or water increased the output of a worker by a factor of about 1000, due to the application of James Hargreaves' spinning jenny, Richard Arkwright's water frame, Samuel Crompton's Spinning Mule and other inventions. The power loom increased the output of a worker by a factor of over 40. The cotton gin increased productivity or removing seed from cotton by a factor of 50. Large gains in productivity also occurred in spinning and weaving of wool and linen, but they were not as great as in cotton.

Finance

The Rothschild family revolutionized international finance. The Frankfurt terminus of the Taunus Railway was financed by the Rothschild family and opened in 1840 as one of Germany's first railways.

The growth of Britain's industry stimulated a concomitant growth in her system of finance and credit. In the 18th century, services offered by banks increased. Clearing facilities, security investments, checks and overdraft protections were introduced. Checks were invented in the 17th century in England and banks settled payments by direct courier to the issuing bank. Around 1770, they began meeting in a central location, and by the 19th century a dedicated space was established, known as a bankers' clearing house. The London clearing house used a method where each bank paid cash to and then was paid cash by an inspector at the end of each day. The first overdraft facility was set up in 1728 by The Royal Bank of Scotland.

The end of the Napoleonic War and the subsequent rebound in trade, led to an expansion in the bullion reserves held by the Bank of England, from a low of under 4 million pounds in 1821 to 14 million pounds by late 1824.

Older innovations became routine parts of financial life during the 19th-century. The Bank of England first issued bank notes during the 17th century, but the notes were hand written and few in number. After 1725 they were partially printed, but cashiers still had to sign each note and make them payable to a named person. In 1844, parliament passed the Bank Charter Act tying these notes to gold reserves, effectively creating the institution of central banking and monetary policy. The notes became fully printed and widely available from 1855.

Growing international trade increased the number of banks, especially in London. These new "merchant banks" facilitated trade growth, profiting from England's emerging dominance in seaborne shipping. Two immigrant families, Rothschild and Baring, established merchant banking firms in London in the late 18th century and came to dominate world banking in the next century. The tremendous wealth amassed by these banking firms soon attracted much attention. The poet George Gordon Byron wrote in 1823: "Who makes politics run glibber all?/ The shade of Bonaparte's noble daring?/ Jew Rothschild and his fellow-Christian, Baring."

The operation of banks also shifted. At the beginning of the century, banking was still an elite preoccupation of a handful of very wealthy families. Within a few decades, however, a new sort of banking had emerged, owned by anonymous stockholders, run by professional managers, and the recipient of the deposits of a growing body of small middle-class savers. Although this new breed of banks was new in prominence, it was not newly invented – the Quaker family Barclays, had been banking in this manner since 1690.

Free trade and globalization

At the height of the First French Empire, Napoleon sought to introduce a "continental system" that would render Europe economically autonomous, thereby emasculating British trade and commerce. It involved such stratagems as the use of beet sugar in preference to the cane sugar that had to be imported from the tropics. Although this caused businessmen in England to agitate for peace, the government was able to persevere, in part because the United Kingdom was well into the industrial revolution. The war had the opposite effect - it stimulated the growth of certain industries like pig-iron output, which was just 68,000 tons in 1788 and soared to 244,000 tons by 1806.

19th century Great Britain become the first global economic superpower, because of superior manufacturing technology and improved global communications such as steamships and railroads.

In 1817, David Ricardo, James Mill and Robert Torrens showed that free trade would benefit the industrially weak as well as the strong, in the famous theory of comparative advantage. In Principles of Political Economy and Taxation Ricardo advanced the doctrine still considered the most counterintuitive in economics:

When an inefficient producer sends the merchandise it produces best to a country able to produce it more efficiently, both countries benefit.

By the mid 19th century, Britain was firmly wedded to the notion of free trade and the first era of globalization began. In the 1840s, the Corn Laws and the Navigation Acts were repealed, ushering in a new age of free trade. In line with the teachings of the classical political economists, led by Adam Smith and David Ricardo, Britain embraced liberalism, encouraging competition and the development of a market economy.

Industrialization allowed cheap production of household items using economies of scale, while rapid population growth created sustained demand for commodities. Globalization in this period was decisively shaped by nineteenth-century imperialism. After the First and Second Opium Wars and the completion of British conquest of India, vast populations of these regions became ready consumers of European exports. It was in this period that areas of sub-Saharan Africa and the Pacific islands were incorporated into the world system. Meanwhile, the conquest of new parts of the globe, notably sub-Saharan Africa, by Europeans yielded valuable natural resources such as rubber, diamonds and coal and helped fuel trade and investment between the European imperial powers, their colonies, and the United States.

The gold standard formed the financial basis of the international economy from 1870–1914.

"The inhabitant of London could order by telephone, sipping his morning tea, the various products of the whole earth, and reasonably expect their early delivery upon his doorstep. Militarism and imperialism of racial and cultural rivalries were little more than the amusements of his daily newspaper. What an extraordinary episode in the economic progress of man was that age which came to an end in August 1914. "

The global financial system was mainly tied to the gold standard in this period. The United Kingdom first formally adopted this standard in 1821. Soon to follow was Canada in 1853, Newfoundland in 1865, and the United States and Germany (de jure) in 1873. New technologies, such as the telegraph, the transatlantic cable, the Radiotelephone, the steamship, and railway allowed goods and information to move around the world at an unprecedented degree.

The eruption of civil war in the United States in 1861 and the blockade of its ports to international commerce meant that the main supply of cotton for the Lancashire looms was cut off. The textile industries shifted to reliance upon cotton from Africa and Asia during the course of the U.S. civil war, and this fact created pressure for an Anglo-French controlled canal through the Suez peninsula. The Suez Canal opened in 1869; in the same year the Central Pacific Railroad that spanned the North American continent was completed.

Capitalism and the engine of profit were making the globe a smaller place.

Twentieth century

Several major challenges to capitalism appeared in the early part of the 20th century. The Russian revolution in 1917 established the first socialist state in the world; a decade later, the Great Depression triggered increasing criticism of the existing capitalist system. One response to this crisis was a turn to fascism, an ideology which advocated state-influenced capitalism; whilst others rejected capitalism altogether in favor of communist or socialist ideologies.

Keynesianism and free markets

The economic recovery of the world's leading capitalist economies in the period following the end of the Great Depression and the Second World War—a period of unusually rapid growth by historical standards—eased discussion of capitalism's eventual decline or demise. The state began to play an increasingly prominent role to moderate and regulate the capitalistic system throughout much of the world.

Keynesian economics became a widely accepted method of government regulation and countries such as the United Kingdom experimented with mixed economies in which the state owned and operated certain major industries.

The state also expanded in the US; in 1929, total government expenditures amounted to less than one-tenth of GNP; from the 1970s they amounted to around one-third. Similar increases were seen in all industrialized capitalist economies, some of which, such as France, have reached even higher ratios of government expenditures to GNP than the United States.

A broad array of new analytical tools in the social science field were developed to explain the social and economic trends of the period including the concepts of post-industrial society and the welfare state.

The long postwar boom ended in the 1970s, amid the economic crises experienced following the 1973 oil crisis. The "stagflation" of the 1970s led many economic commentators and politicians to embrace market-oriented policy prescriptions inspired by the laissez-faire capitalism and classical liberalism of the 19th century, particularly under the influence of Friedrich Hayek and Milton Friedman. The theoretical alternative to Keynesianism was more compatible with laissez-faire and emphasized rapid expansion of the economy. Market-oriented solutions gained increasing support in the capitalist world, especially under leadership of Ronald Reagan in the U.S. and Margaret Thatcher in the UK in the 1980s. Public and political interest began shifting away from the so-called collectivist concerns of Keynes's managed capitalism to a focus on individual choice, called "remarketized capitalism." In the eyes of many economic and political commentators, the collapse of the Soviet Union brought further evidence of the superiority of market capitalism over planned economy.

Globalization

Although overseas trade has been associated with the development of capitalism for over five hundred years, some thinkers argue that a number of trends associated with globalization have acted to increase the mobility of people and capital since the last quarter of the 20th century, combining to circumscribe the room to maneuver of states in choosing non-capitalist models of development. Today, these trends have bolstered the argument that capitalism should now be viewed as a truly world system (Burnham). However, other thinkers argue that globalization, even in its quantitative degree, is no greater now than during earlier periods of capitalist trade.

After the abandonment of the Bretton Woods system in 1971, and the strict state control of foreign exchange rates, the total value of transactions in foreign exchange was estimated to be at least twenty times greater than that of all foreign movements of goods and services (EB). The internationalization of finance, which some see as beyond the reach of state control, combined with the growing ease with which large corporations have been able to relocate their operations to low-wage states, has posed the question of the 'eclipse' of state sovereignty, arising from the growing 'globalization' of capital.

While economists generally agree about the size of global income inequality, there is a general disagreement about the recent direction of change of it. In cases such as China, where income inequality is clearly growing it is also evident that overall economic growth has rapidly increased with capitalist reforms. The book The Improving State of the World, published by the libertarian think tank Cato Institute, argues that economic growth since the Industrial Revolution has been very strong and that factors such as adequate nutrition, life expectancy, infant mortality, literacy, prevalence of child labor, education, and available free time have improved greatly. Some scholars, including Stephen Hawking and researchers for the International Monetary Fund, contend that globalization and neoliberal economic policies are not ameliorating inequality and poverty but exacerbating it, and are expanding populations of the displaced, the unemployed and the imprisoned along with accelerating the destruction of the environment and species extinction.

Today

By the beginning of the 21st century, capitalism had become the pervasive economic system worldwide. The collapse of the Soviet bloc in 1991 significantly reduced the influence of Socialism as an alternative economic system. Socialist movements continue to be influential in some parts of the world, most notably Latin-American Bolivarianism, with some having ties to more traditional anti-capitalist movements, such as Bolivarian Venezuela's ties to communist Cuba.

In many emerging markets, the influence of banking and financial capital have come to increasingly shape national developmental strategies, leading some to argue we are in a new phase of financial capitalism.

State intervention in global capital markets following the financial crisis of 2007–2010 was perceived by some as signaling a crisis for free-market capitalism. Serious turmoil in the banking system and financial markets due in part to the subprime mortgage crisis reached a critical stage during September 2008, characterized by severely contracted liquidity in the global credit markets posed an existential threat to investment banks and other institutions.

Future

According to some, the transition to the information society involves abandoning some parts of capitalism, as the "capital" required to produce and process information becomes available to the masses and difficult to control, and is closely related to the controversial issues of intellectual property. Some even speculate that the development of mature nanotechnology, particularly of universal assemblers, may make capitalism obsolete, with capital ceasing to be an important factor in the economic life of humanity.

Appendix 4

The Target Human Population Count

How can humans achieve zero ecological footprint? We will try to do that via two approaches. One approach is to take the sun's energy allotment for humans and calculate the number of humans who should be living within the Earth's boundaries. We will call this the top-down method.

The second approach is to take a stable long-lasting human population that exists today and extrapolate from it the number of humans that can be supported on Earth. We will call this the bottom-up method.

Top-Down Analysis

Rank	Category	# Individuals (Billions)
1	Blue whales	0
2	Cattle	1
3	Sheep & Goats	2
4	Humans	7
5	Chickens	24
6	Ants	10,000,000
7	Antartic crill	800,000,000,000,000
8	Bacteria	5,000,000,000,000,000,000,000,000,000,000

A table of Earth life forms ranked by individual counts

Rank	Category	Biomass (Wet million tones)
1	Blue whales	36
2	Chickens	48
3	Sheep & Goats	105
4	Humans	350
5	Antartic crill	379
6	Cattle	520
7	Picoplankton	1,000
8	Ants	5,000
9	Bacteria	680,000

A table of Earth life forms ranked by biomass

To reduce the ecological footprint of humans to zero, we need to talk about the human population, and its allotment of the energy from the Sun.

It must be realized that the rays of the sun are what makes the whole Earth come alive and the amount of sunlight coming to Earth is approximately 10,547,945,205,479,500,000,000 joules/day where only 71% is absorbed by the planet which is equal to 7,489,041,095,890,410,000,000 joules/day. Please note this figure represents both land-based and ocean-based rays. (1)

Plants are converters of this energy into plant matter and their average photosynthetic efficiency is approximately 4.5%. That leaves us with 337,006,849,315,069,000,000 joules/day for all the living on Earth (2).

The biomass of present humans on Earth is approximately 0.0509396% of all the living creatures. This figure represents approximately 7.4 billion individuals which many experts consider too many to be sustainable. I will use this percentage as a starting point and consider it in the sum of all human biomass plus the biomass of all its food resources (3).

Therefore, the basic quota of energy for consumption by humans is 171,670,000,437,024,000 joules/day (4).

Given that

A) The percentage of animal matter in the human diet is 5.0%

B) The percentage of plant matter in the human diet is 95.0%

C) There are 4.184 joules in a calorie

D) Calorie intake of a human is approximately 1,800,000/day (The calories count you and I used in our daily lives is 1,800/day)

E) Humans consume 5% of growing plant matter (ratio of the whole plant to the grain is 20:1)

F) The ratio of plant matter required to produce edible animal matter is 2.5% (ratio of plants to edible meat is 40:1)

Energy consumption per human is at 753,120,000 joules/day (5).

Given that the human energy consumption efficiency is 20% (6), the total number of humans that can be supported based on their proportion of the total Earth biota is approximately 45,589,016 individuals (7).

The total land mass of arable land that supports the human population is 15 million kilometers squared. Therefore, each human on average occupies and gains resources from approximately 330,000 meters squared of arable land. (8)

Bottom-Up Analysis

The BaMbuti people live in the Ituri Forest, a tropical rainforest covering approximately 63,000 km^2 of the north/northeast section of the Democratic Republic of the Congo, Africa. BaMbuti are pygmy hunter-gatherers, and are some of the oldest indigenous people of the Congo region. They are composed of bands ranging from 15 to 60 people, with a population hovering around 30,000 to 40,000. It is not known how long they, the pigmy people, have been living in the Ituri forest, but there have been reports of their existence since the pharaohs of Egypt approximately 4,000 years ago (*The Forest People* by Turnbull, Colin M.).

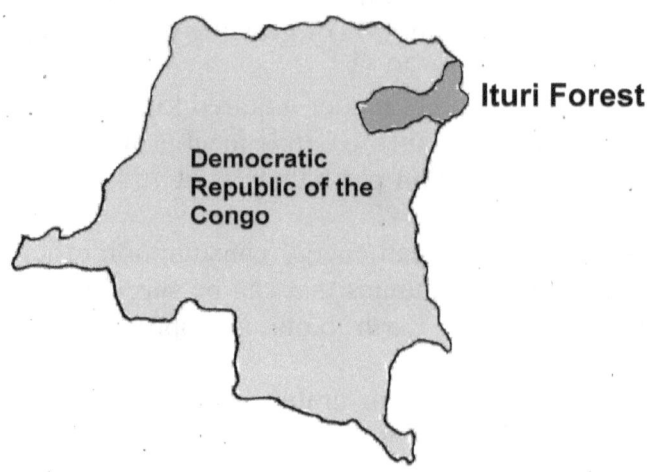

Location of the Ituri Forest in the Democratic Republic of the Congo

The Ituri Forest receives a large amount of rainfall annually, ranging from 50 to 70 inches. The dry season is relatively short, ranging from 1 to 2 months. The forest is a moist, humid region strewn with rivers and lakes. Too much rainfall, as well as droughts, can greatly diminish the food supply.

In the rainy season, pigmy people forage in a part of the forest, and in the dry season, they expand their foraging territories. Their wild foodstuffs include crabs, shellfish, ants, larvae, snails, pigs, antelope, monkeys, and fish as well as honey. The plant component of their diet includes wild yams, berries, fruits, roots, leaves and cola nuts. They also eat mushrooms.

The habitat area for each human in the Ituri forest is approximately 1,500,000 meters squared. Because they have been living in the forest for thousands of years with zero ecological footprint, we can try to extrapolate the population size if the whole Earth lived as they do; the population of the Earth should be approximately 10 million individuals (9).

As you recall from the top-down method discussed previously, the assumed population size was approximately 46 million individuals. It is argued here that these figures are close to one another and may represent the range of the Earth's population that can exist at zero ecological footprint.

Reading List

This reading list covers the major topics in the current and future events in human history.

2052: A Global Forecast for the Next Forty Years Paperback; Jorgen Randers; June 13, 2012

Beyond the Inflection Point: An Economic Defense of the Limits-To-Growth Theory; Andrew Currie

Cannibal Business: The Limits to Growth; Bruce Ferguson

Feeding the Fire: The Lost History and Uncertain Future of Mankind's Energy Addiction; Mark Eberhart; May 8, 2007

Limits to Growth: The 30-Year Update; Donella H. Meadows and Jorgen Randers; Jun 1, 2004

Scatter, Adapt, and Remember: How Humans Will Survive a Mass Extinction; Annalee Newitz

The End of Growth: Adapting to Our New Economic Reality; August 9, 2011

The No-Growth Imperative: Creating Sustainable Communities under Ecological Limits to Growth; Gabor Zovanyi

The Sixth Extinction: An Unnatural History; Elizabeth Kolbert; January 6, 2015

Under a Green Sky: Global Warming, the Mass Extinctions of the Past, and What They Can Tell Us About Our Future; Peter D. Ward; March 25, 2008

P.R. Brewbaker

www.ingramcontent.com/pod-product-compliance
Lightning Source LLC
Chambersburg PA
CBHW022000170526
45157CB00003B/1083